Pitman Research Notes in Mathematics Series

Main Editors
H. Brezis, Université de Paris
R. G. Douglas, State University of New York at Stony Brook
A. Jeffrey, University of Newcastle-upon-Tyne *(Founding Editor)*

Editorial Board
R. Aris, University of Minnesota
A. Bensoussan, INRIA, France
W. Bürger, Universität Karlsruhe
J. Douglas Jr, University of Chicago
R. J. Elliott, University of Hull
G. Fichera, Università di Roma
R. P. Gilbert, University of Delaware
R. Glowinski, Université de Paris
K. P. Hadeler, Universität Tübingen
K. Kirchgässner, Universität Stuttgart
B. Lawson, State University of New York at Stony Brook
W. F. Lucas, Cornell University
R. E. Meyer, University of Wisconsin-Madison
J. Nitsche, Universität Freiburg
L. E. Payne, Cornell University
G. F. Roach, University of Strathclyde
J. H. Seinfeld, California Institute of Technology
I. N. Stewart, University of Warwick
S. J. Taylor, University of Virginia

Submission of proposals for consideration
Suggestions for publication, in the form of outlines and representative samples, are invited by the Editorial Board for assessment. Intending authors should approach one of the main editors or another member of the Editorial Board, citing the relevant AMS subject classifications. Alternatively, outlines may be sent directly to the publisher's offices. Refereeing is by members of the board and other mathematical authorities in the topic concerned, throughout the world.

Preparation of accepted manuscripts
On acceptance of a proposal, the publisher will supply full instructions for the preparation of manuscripts in a form suitable for direct photo-lithographic reproduction. Specially printed grid sheets are provided and a contribution is offered by the publisher towards the cost of typing. Word processor output, subject to the publisher's approval, is also acceptable.

Illustrations should be prepared by the authors, ready for direct reproduction without further improvement. The use of hand-drawn symbols should be avoided wherever possible, in order to maintain maximum clarity of the text.

The publisher will be pleased to give any guidance necessary during the preparation of a typescript, and will be happy to answer any queries.

Important note
In order to avoid later retyping, intending authors are strongly urged not to begin final preparation of a typescript before receiving the publisher's guidelines and special paper. In this way it is hoped to preserve the uniform appearance of the series.

Longman Scientific & Technical
Longman House
Burnt Mill
Harlow, Essex, UK
(tel (0279) 26721)

Titles in this series

54 Geometry of Coxeter groups
H Hiller
55 Cusps of Gauss mappings
T Banchoff, T Gaffney and C McCrory
56 An approach to algebraic K-theory
A J Berrick
57 Convex analysis and optimization
J-P Aubin and R B Vintner
58 Convex analysis with applications in the differentiation of convex functions
J R Giles
59 Weak and variational methods for moving boundary problems
C M Elliott and J R Ockendon
60 Nonlinear partial differential equations and their applications: Collège de France Seminar. Volume II
H Brezis and J L Lions
61 Singular systems of differential equations II
S L Campbell
62 Rates of convergence in the central limit theorem
Peter Hall
63 Solution of differential equations by means of one-parameter groups
J M Hill
64 Hankel operators on Hilbert space
S C Power
65 Schrödinger-type operators with continuous spectra
M S P Eastham and H Kalf
66 Recent applications of generalized inverses
S L Campbell
67 Riesz and Fredholm theory in Banach algebra
B A Barnes, G J Murphy, M R F Smyth and T T West
68 Evolution equations and their applications
F Kappel and W Schappacher
69 Generalized solutions of Hamilton-Jacobi equations
P L Lions
70 Nonlinear partial differential equations and their applications: Collège de France Seminar. Volume III
H Brezis and J L Lions
71 Spectral theory and wave operators for the Schrödinger equation
A M Berthier
72 Approximation of Hilbert space operators I
D A Herrero
73 Vector valued Nevanlinna Theory
H J W Ziegler
74 Instability, nonexistence and weighted energy methods in fluid dynamics and related theories
B Straughan
75 Local bifurcation and symmetry
A Vanderbauwhede
76 Clifford analysis
F Brackx, R Delanghe and F Sommen
77 Nonlinear equivalence, reduction of PDEs to ODEs and fast convergent numerical methods
E E Rosinger
78 Free boundary problems, theory and applications. Volume I
A Fasano and M Primicerio
79 Free boundary problems, theory and applications. Volume II
A Fasano and M Primicerio
80 Symplectic geometry
A Crumeyrolle and J Grifone
81 An algorithmic analysis of a communication model with retransmission of flawed messages
D M Lucantoni
82 Geometric games and their applications
W H Ruckle
83 Additive groups of rings
S Feigelstock
84 Nonlinear partial differential equations and their applications: Collège de France Seminar. Volume IV
H Brezis and J L Lions
85 Multiplicative functionals on topological algebras
T Husain
86 Hamilton-Jacobi equations in Hilbert spaces
V Barbu and G Da Prato
87 Harmonic maps with symmetry, harmonic morphisms and deformations of metrics
P Baird
88 Similarity solutions of nonlinear partial differential equations
L Dresner
89 Contributions to nonlinear partial differential equations
C Bardos, A Damlamian, J I Díaz and J Hernández
90 Banach and Hilbert spaces of vector-valued functions
J Burbea and P Masani
91 Control and observation of neutral systems
D Salamon
92 Banach bundles, Banach modules and automorphisms of C^*-algebras
M J Dupré and R M Gillette
93 Nonlinear partial differential equations and their applications: Collège de France Seminar. Volume V
H Brezis and J L Lions
94 Computer algebra in applied mathematics: an introduction to MACSYMA
R H Rand
95 Advances in nonlinear waves. Volume I
L Debnath
96 FC-groups
M J Tomkinson
97 Topics in relaxation and ellipsoidal methods
M Akgül
98 Analogue of the group algebra for topological semigroups
H Dzinotyiweyi
99 Stochastic functional differential equations
S E A Mohammed
100 Optimal control of variational inequalities
V Barbu
101 Partial differential equations and dynamical systems
W E Fitzgibbon III
102 Approximation of Hilbert space operators. Volume II
C Apostol, L A Fialkow, D A Herrero and D Voiculescu
103 Nondiscrete induction and iterative processes
V Ptak and F-A Potra

104 Analytic functions – growth aspects
O P Juneja and G P Kapoor
105 Theory of Tikhonov regularization for Fredholm equations of the first kind
C W Groetsch
106 Nonlinear partial differential equations and free boundaries. Volume I
J I Díaz
107 Tight and taut immersions of manifolds
T E Cecil and P J Ryan
108 A layering method for viscous, incompressible L_p flows occupying R^n
A Douglis and E B Fabes
109 Nonlinear partial differential equations and their applications: Collège de France Seminar. Volume VI
H Brezis and J L Lions
110 Finite generalized quadrangles
S E Payne and J A Thas
111 Advances in nonlinear waves. Volume II
L Debnath
112 Topics in several complex variables
E Ramírez de Arellano and D Sundararaman
113 Differential equations, flow invariance and applications
N H Pavel
114 Geometrical combinatorics
F C Holroyd and R J Wilson
115 Generators of strongly continuous semigroups
J A van Casteren
116 Growth of algebras and Gelfand–Kirillov dimension
G R Krause and T H Lenagan
117 Theory of bases and cones
P K Kamthan and M Gupta
118 Linear groups and permutations
A R Camina and E A Whelan
119 General Wiener–Hopf factorization methods
F-O Speck
120 Free boundary problems: applications and theory, Volume III
A Bossavit, A Damlamian and M Fremond
121 Free boundary problems: applications and theory, Volume IV
A Bossavit, A Damlamian and M Fremond
122 Nonlinear partial differential equations and their applications: Collège de France Seminar. Volume VII
H Brezis and J L Lions
123 Geometric methods in operator algebras
H Araki and E G Effros
124 Infinite dimensional analysis–stochastic processes
S Albeverio
125 Ennio de Giorgi Colloquium
P Krée
126 Almost-periodic functions in abstract spaces
S Zaidman
127 Nonlinear variational problems
A Marino, L Modica, S Spagnolo and M Degiovanni
128 Second-order systems of partial differential equations in the plane
L K Hua, W Lin and C-Q Wu
129 Asymptotics of high-order ordinary differential equations
R B Paris and A D Wood
130 Stochastic differential equations
R Wu
131 Differential geometry
L A Cordero
132 Nonlinear differential equations
J K Hale and P Martinez-Amores
133 Approximation theory and applications
S P Singh
134 Near-rings and their links with groups
J D P Meldrum
135 Estimating eigenvalues with *a posteriori/a priori* inequalities
J R Kuttler and V G Sigillito
136 Regular semigroups as extensions
F J Pastijn and M Petrich
137 Representations of rank one Lie groups
D H Collingwood
138 Fractional calculus
G F Roach and A C McBride
139 Hamilton's principle in continuum mechanics
A Bedford
140 Numerical analysis
D F Griffiths and G A Watson
141 Semigroups, theory and applications. Volume I
H Brezis, M G Crandall and F Kappel
142 Distribution theorems of L-functions
D Joyner
143 Recent developments in structured continua
D De Kee and P Kaloni
144 Functional analysis and two-point differential operators
J Locker
145 Numerical methods for partial differential equations
S I Hariharan and T H Moulden
146 Completely bounded maps and dilations
V I Paulsen
147 Harmonic analysis on the Heisenberg nilpotent Lie group
W Schempp

Harmonic analysis on the Heisenberg nilpotent Lie group

W Schempp
University of Siegen

Harmonic analysis on the Heisenberg nilpotent Lie group, with applications to signal theory

Longman Scientific & Technical
Copublished in the United States with
John Wiley & Sons, Inc., New York

Longman Scientific & Technical
Longman Group UK Limited
Longman House, Burnt Mill, Harlow
Essex CM20 2JE, England
and Associated Companies throughout the world.

Copublished in the United States with
John Wiley & Sons, Inc., 605 Third Avenue, New York, NY 10158

First published 1986

AMS Subject Classifications: (main) 22E27, 43A35, 94A12
(subsidiary) 22D10, 22E25, 41A15

ISSN 0269-3674

British Library Cataloguing in Publication Data
Schempp, W.
Harmonic analysis on the Heisenberg
nilpotent Lie group, with applications to
signal theory.—(Pitman research notes
in mathematics, ISSN 0269-3674; 147)
1. Lie groups, Nilpotent
I. Title
512′.55 QA387

ISBN 0-582-99453-5

Library of Congress Cataloging-in-Publication Data
Schempp. W. (Walter), 1938–
Harmonic analysis on the Heisenberg nilpotent Lie
group, with applications to signal theory.

(Pitman research notes in mathematics series,
ISSN 0269-3674; 147)
Bibliography: p.
Includes index.
1. Harmonic analysis. 2. Lie groups, Nilpotent.
3. Signal theory (Telecommunication) I. Title.
II. Title: Heisenberg nilpotent Lie group, with
applications to signal theory. III. Series: Pitman
research notes in mathematics; 147.
QA403.S27 1986 515′.2433 86-13233
ISBN 0-470-20374-9 (USA only)

Printed and bound in Great Britain by
Biddles Ltd, Guildford and King's Lynn

Contents

Preface

The real Heisenberg group $\tilde{A}(\mathbb{R})$ is a connected and simply connected, two-step nilpotent, analytic group having one-dimensional centre $\tilde{C}$. Therefore $\tilde{A}(\mathbb{R})$ forms the simplest possible non-commutative, non-compact Lie group. The name and the quantum mechanical meaning of the real Heisenberg nilpotent Lie group $\tilde{A}(\mathbb{R})$ stem from the fact that the Lie algebra $\mathfrak{n}$ of $\tilde{A}(\mathbb{R})$ over $\mathbb{R}$ is defined by the Heisenberg canonical commutation relations. Thus, according to the philosophy of Niels Bohr, the geometric intuition necessarily fails to describe the action of $\tilde{A}(\mathbb{R})$. It is the purpose of these notes to study nilpotent harmonic analysis in a unified manner and specifically to determine the unitary dual of $\tilde{A}(\mathbb{R})$ by an application of the Mackey machinery as well as by the Kirillov orbit picture. The coadjoint orbit method provides a deep geometric insight into the harmonic analysis of the Heisenberg Lie group. Although the unitary dual of $\tilde{A}(\mathbb{R})$ is extremely poor, there are many rather different looking ways of realizing the non-degenerate, topologically irreducible, continuous, unitary, linear representations of $\tilde{A}(\mathbb{R})$. This important fact adds greatly to the applicability of the real Heisenberg nilpotent Lie group $\tilde{A}(\mathbb{R})$ and turns it into a far reaching tool for various different areas of pure and applied mathematics, theoretical physics, information theory, and electrical engineering. In the present notes, however, the main emphasis of the applications are laid on the theory of analog and digital signals since the group theoretical ideas behind this subject have been discovered quite recently. Specifically the notes present the solutions of two problems of analog radar signal design: the synthesis problem of characterizing intrinsically the bivariate analog radar auto-ambiguity functions and the invariant problem of computational signal geometry of calculating explicitly the linear energy preserving automorphisms of the radar ambiguity surfaces over the symplectic time-frequency plane. Both the solutions are achieved via harmonic analysis on the differential principal fibre bundle over the two-dimensional polarized resp. isotropic cross-section with structure group isomorphic to the centre $\tilde{C}$ of $\tilde{A}(\mathbb{R})$. Moreover it it shown how the linear lattice representation of $\tilde{A}(\mathbb{R})$ gives rise to a geometric proof of the sampling theorem of digital signal processing and how to deduce basically from the preceding results some new identities for Laguerre functions and theta-null values.

Those parts of the notes which are concerned with elementary group representation theory are based on lectures entitled "Einführung in die Darstellungstheorie lokalkompakter topologischer Gruppen" given by the author at the University of Siegen in Winter Semester 1983/84. Apart from the theory of analog and digital signals there are, however, various other applications of harmonic analysis on the principal differential fibre bundle over the two-dimensional polarized resp. isotropic cross-section with structure group isomorphic to the centre $\tilde{C}$ of $\tilde{A}(\mathbb{R})$, and the closely related theory of the Segal-Shale-Weil metaplectic (or linear oscillator) representation for reductive dual pairs in metaplectic groups. Among the applications which are of actual interest from the technological point of view we should particularly emphasize the field of beam optics, the synthesis for dielectric multilayer filters, the theory of transmission by dielectric waveguides and optical distributed-index round fibres, the design of maser and laser resonators including optical phase conjugation devices, and the holographic imaging. Unfortunately a detailed treatment of these topics and their applications to optical communication systems of high capacity are outside the scope of the present notes. However these notes will form the foundation of a research program dealing with a group representational approach to certain optical phenomena. A series of invited papers appearing in the near future will trace a research line which follows Charles H. Townes' example by starting from analog radar signal design and leading via group theoretical arguments to a detailed treatment of various devices of microwave and laser optics.

Walter Schempp

Lehrstuhl fuer Mathematik I
University of Siegen

Acknowledgements

First and foremost the author wishes to acknowledge the invaluable stimulation and encouragement supplied by Professors Aline Bonami (Orléans), Chen Han-lin (Beijing), Charles K. Chui (College Station), Lothar Collatz (Hamburg), Phillipe Combe (Marseille), Rudolf de Buda (Toronto), Walter Gautschi (West Lafayette), Michiel Hazewinkel (Amsterdam), Edwin Hewitt (Seattle), J. Rowland Higgins (Cambridge), Hu Ying-sheng (Beijing), Mourad E.H. Ismail (Tempe), Palle E.T. Jørgensen (Iowa City), John R. Klauder (Murray Hill), Adam Korányi (Bronx), Peter Kramer (Tübingen), Giovanni Monegato (Torino), Günter Ries (Siegen), Rudolf Schwarte (Siegen), Harvey A. Smith (Tempe), Orestes N. Stavroudis (Tucson), Niels Henrik Stetkaer (Århus), Daniele C. Struppa (Pisa), Kurt Bernardo Wolf (Mexico City), and the microwave engineer Jerzy Brzeski (Walnut Creek).

Moreover, the author wishes to thank the various Universities, Research Institutes, and funding agencies which have offered research facilities, hospitality, and support during the conduct of the research parts of which are reported here. Specifically, the author's thanks go to Aarhus Universitet at Århus, Denmark; Academia Sinica at Beijing, The People's Republic of China; Arizona State University at Tempe, Arizona; the Bulgarian Academy of Sciences at Sofia, Bulgaria; the Center for Approximation Theory at College Station, Texas; to Centro de Investigaciones en Optica at Léon, Guanajuato, México; to the Centrum voor Wiskunde en Informatica at Amsterdam, The Netherlands; to the Mathematisches Forschungsinstitut Oberwolfach, Black Forest, Germany; to the Scuola Normale Superiore at Pisa, Italy ; to the University of Alaska at Fairbanks, Alaska; University of Delaware at Newark, Delaware; University of Illinois, Urbana-Champaign, Illinois; University of Maryland, College Park, Maryland; Universidad Nacional Autónoma de México at México City: Université d'Orléans at Orléans, France; Pennsylvania State University at University Park, Pennsylvania; Seoul National University in Seoul, Korea; Università di Torino in Turin, Italy; University of Washington at Seattle, Washington, Zentrum fuer interdisziplinaere Forschung (ZiF) at Bielefeld, Germany; Consiglio Nazionale delle Ricerche of Italy, Deutscher Akademischer

Austauschdienst, Deutsche Forschungsgemeinschaft, Korean Physical Society, Scuola Matematica Interuniversitaria, and finally, to the National Science Foundation.

0 Basic notations and conventions

First, the symbols $\mathbb{N}$, $\mathbb{Z}$, $\mathbb{Q}$, $\mathbb{R}$, and $\mathbb{C}$ denote the sets of natural, integral, rational, real, and complex numbers, respectively: $\mathbb{R}^\times$ denotes the set of real numbers $\neq 0$ and $\mathbb{T}$ the set of complex numbers of modulus 1. Throughout these notes we shall adhere to the following conventions. By the terms 'locally compact topological group' and 'Lie group' we shall understand a locally compact topological group and a Lie group that are countable at infinity. A simply connected Lie group shall mean a connected and simply connected Lie group. Finally, by the term 'Hilbert space' we shall mean a separable Hilbert space.

At the end of each chapter there is a list of references which supply additional material. Special emphasis is laid on survey articles which include further references.

1 Basic facts on linear group representations

1.1 Let G denote a group. Write its group law in the multiplicative way and denote by 1_G the neutral element of G. A *linear representation* of the group G in a complex vector space $\mathcal{H}$ is a pair $(U,\mathcal{H})$ where U denotes a mapping which assigns to every element $x \in G$ a $\mathbb{C}$-linear mapping $U(x) : \mathcal{H} \to \mathcal{H}$ such that the following two conditions are satisfied:

(I) $U(1_G) = \mathrm{id}_{\mathcal{H}}$ (the identity operator of $\mathcal{H}$);

(II) For all pairs $(x,y) \in G \times G$ we have $U(xy) = U(x) \circ U(y)$.

Obviously we have $U(x) \circ U(x^{-1}) = \mathrm{id}_{\mathcal{H}}$ and therefore $U(x^{-1}) = (U(x))^{-1}$ for all $x \in G$. Thus a linear representation $(U,\mathcal{H})$ of G in the *representation space* $\mathcal{H}$ defines a *morphism* $x \rightsquigarrow U(x)$ of the group G into the group $\mathrm{Aut}(\mathcal{H}) = \underline{\underline{GL}}(\mathcal{H})$ of automorphisms of $\mathcal{H}$. In the case when G is a topological group and $\mathcal{H}$ denotes a topological vector space over the field $\mathbb{C}$, a linear representation $(U,\mathcal{H})$ of G is said to be *continuous* if the linear left G-action on $\mathcal{H}$ canonically defined by the assignment

$$G \times \mathcal{H} \ni (x,f) \rightsquigarrow U(x)f \in \mathcal{H}$$

is a continuous mapping with respect to the product topology of $G \times \mathcal{H}$ and the given vector space topology of $\mathcal{H}$. Thus the *left* G-*module* $\mathcal{H}$ becomes a *topological* left G-module. More specifically, when $(\mathcal{H};\langle .|.\rangle)$ is a complex Hilbert space with norm $\|\cdot\|$ associated with its scalar product $\langle .|.\rangle$, a linear representation $(U,\mathcal{H})$ of G is said to be *unitary* if the automorphism $U(x) \in \underline{\underline{GL}}(\mathcal{H})$ forms a unitary operator of $\mathcal{H}$ for all $x \in G$. In this case U defines a morphism $x \rightsquigarrow U(x)$ of the group G into the unitary group $\underline{\underline{U}}(\mathcal{H})$ of $\mathcal{H}$ such that $U(x^{-1}) = U(x)^*$ holds for all $x \in G$. Clearly the unitarity of the linear representation $(U,\mathcal{H})$ of G is equivalent to one of the following two equivalent conditions:

(i) $\|U(x)f\| = \|f\|$ for all $x \in G$ and all $f \in \mathcal{H}$;

(ii) $\langle U(x)f|U(x)g\rangle = \langle f|g\rangle$ for all $x \in G$ and all pairs $(f,g) \in \mathcal{H} \times \mathcal{H}$.

In the case when $(U,\mathcal{H})$ is a unitary linear representation of the topological group G in the complex Hilbert space $\mathcal{H}$ the property of being continuous as defined previously follows if $(U,\mathcal{H})$ is merely supposed to be *separately continuous*, to wit, if the mapping given by

$$G \ni x \rightsquigarrow U(x)f \in \mathcal{H}$$

is supposed to be continuous for any choice of the vector $f \in \mathcal{H}$. Indeed, let $x_0 \in G$ and $f_0 \in \mathcal{H}$ be fixed elements. For any given $\varepsilon > 0$ there exists a suitable neighbourhood V_0 of x_0 in G such that

$$\|U(x)f_0 - U(x_0)f_0\| < \frac{1}{2}\varepsilon$$

holds for all $x \in V_0$. Let the vector $f \in \mathcal{H}$ satisfy $\|f - f_0\| < \frac{1}{2}\varepsilon$. By virtue of (i) we get for all elements $x \in V_0$ the estimate

$$\begin{aligned}\|U(x)f - U(x_0)f_0\| &= \|U(x)f - U(x)f_0 + U(x)f_0 - U(x_0)f_0\| \\ &\leqq \|U(x)(f-f_0)\| + \|U(x)f_0 - U(x_0)f_0\| \\ &< \frac{1}{2}\varepsilon + \frac{1}{2}\varepsilon = \varepsilon.\end{aligned}$$

Thus the linear left G-action $(x,f) \rightsquigarrow U(x)f$ associated with $(U,\mathcal{H})$ is continuous in all points $(x_0,f_0) \in G \times \mathcal{H}$ and hence globally continuous on the topological product space $G \times \mathcal{H}$. The topology induced be the *strong* operator topology of the complex vector space $\mathrm{End}(\mathcal{H})$ of *continuous* endomorphism of $\mathcal{H}$ on the unitary group $\underline{\underline{U}}(\mathcal{H})$ coincides with topology induced by the *weak* operator topology of $\mathrm{End}(\mathcal{H})$ on $\underline{\underline{U}}(\mathcal{H})$ and is *compatible* with the group structure of $\underline{\underline{U}}(\mathcal{H})$.

1.2 Theorem. A unitary linear representation $(U,\mathcal{H})$ of the topological group G in the complex Hilbert space $\mathcal{H}$ is continuous if and only if the morphism of topological groups

$$G \ni x \rightsquigarrow U(x) \in \underline{\underline{U}}(\mathcal{H})$$

is continuous.

The *norm* topology on $\mathrm{End}(\mathcal{H})$, i.e., the topology of uniform convergence on the bounded subsets of $\mathcal{H}$, is finer than the strong operator topology on

$End(\mathcal{H})$ which is finer than the weak operator topology on $End(\mathcal{H})$. However, if $\underline{\underline{U}}(\mathcal{H})$ is equipped with the topology induced by the topology of bounded convergence of the complex Banach space $End(\mathcal{H})$ then the mapping $x \rightsquigarrow U(x)$ is *not* necessarily a continuous morphism. Finally it should be observed that for a continuous linear representation $(U,\mathcal{H})$ of G there does not necessarily exist a Hilbert space structure on the complex topological vector space $\mathcal{H}$ such that $(U,\mathcal{H})$ is a continuous, unitary, linear representation of the topological group G.

Remarks.

1. In our applications to signal theory the representation space $\mathcal{H}$ will be the standard Lebesgue space $L^2(\mathbb{R})$ formed by the equivalence classes of square-integrable complex-valued functions on the real line $\mathbb{R}$. In this case the unitary group $\underline{\underline{U}}(L^2(\mathbb{R}))$ of the complex Hilbert space $L^2(\mathbb{R})$ consists of the energy preserving linear transformations of the complex signal envelopes; cf. Section 8 infra.

2. Let G be a group and $\mathcal{H}$ a complex Hilbert space. Assume that there is a mapping $U : G \to \underline{\underline{U}}(\mathcal{H})$ and a mapping c from $G \times G$ to the one-dimensional compact torus group $\mathbb{T} = \mathbb{R}/\mathbb{Z} = \{\zeta \in \mathbb{C} \mid |\zeta| = 1\}$ such that

$$U(xy) = c(x,y)U(x) \circ U(y)$$

is valid for all pairs $(x,y) \in G \times G$. Obviously the normalizing condition $c(1,y) = c(x,1) = 1$ holds. It follows from the associativity in G or, more precisely, from $U((xy)z) = U(x(yz))$, that the identity

$$c(x,y)c(xy,z) = c(x,yz)c(y,z)$$

holds for all elements x,y,z in G. In this case the pair $(U,\mathcal{H})$ is said to be a unitary, *projective*, linear representation or a *ray representation* of the group G in $\mathcal{H}$ and $c: G \times G \to \mathbb{T}$ is said to be the associated 2-*cocycle of* G *in* $\mathbb{T}$. Form the *central extension of* G *by* $\mathbb{T}$ in the following way: Endow the Cartesian product

$$G_c = G \times \mathbb{T}$$

with the law of multiplication

$$(x,\zeta) \cdot (y,\eta) = (x.y, \zeta\eta\,\bar{c}(x,y)).$$

Then G_c is called to be the *Mackey obstruction group* associated with G. The unitary, projective, linear representation $(U,\mathcal{H})$ of G in $\mathcal{H}$ extends to a *true* unitary, linear representation $(U_c,\mathcal{H})$ of G_c in $\mathcal{H}$ which is defined via the prescription

$$U_c : G_c \ni (x,\zeta) \rightsquigarrow \zeta U(x) \in \underline{\underline{U}}(\mathcal{H}).$$

Indeed, we have the morphism property

$$U_c((x,\zeta) . (y,\eta)) = U_c((xy,\zeta\eta\bar{c}(x,y))$$

$$= \zeta\eta\,\bar{c}(x,y)U(xy)$$

$$= U_c((x,\zeta)) \circ U_c((y,\eta)).$$

For instance, if G denotes the additive group $\mathbb{R} \oplus \mathbb{R}$ and $\lambda \neq 0$ a real number, the mapping which acts for all pairs $(x,y) \in \mathbb{R} \oplus \mathbb{R}$ on the continuous complex-valued functions $\psi \in \mathbb{K}(\mathbb{R})$ with *compact* support in $\mathbb{R}$ according to the rule

$$U_\lambda(x,y)\psi(t) = e^{2\pi i\lambda(-yt + \frac{1}{2}xy)}\,\psi(t-x) \qquad (t \in \mathbb{R})$$

extends to a unitary, projective, linear representation of $\mathbb{R} \oplus \mathbb{R}$ in the complex Hilbert space $L^2(\mathbb{R})$. The associated 2-cocycle of $\mathbb{R} \oplus \mathbb{R}$ in $\mathbb{T}$ is given by the function

$$c_\lambda((x,y),(x',y')) = \exp\left(\pi i\lambda \det\begin{pmatrix} x & x' \\ y & y' \end{pmatrix}\right)$$

for $(x,y) \in \mathbb{R} \oplus \mathbb{R}$ and $(x',y') \in \mathbb{R} \oplus \mathbb{R}$.

We shall come back to the notions of unitary, projective, linear group representation and Mackey obstruction group in the connection with the oscillator representation of the metaplectic group $\underline{\underline{Mp}}(1,\mathbb{R})$; cf. Section 7.8 infra. The 2-cocycle c_1 of $\mathbb{R} \oplus \mathbb{R}$ in $\mathbb{T}$ will be used in Section 8.9 to extend the notion of positive definiteness for complex-valued functions on $\mathbb{R} \oplus \mathbb{R}$.

3. Let G denote a real Lie group and $(U,\mathcal{H})$ a continuous, unitary, linear representation of G in the complex Hilbert space $\mathcal{H}$. For $k \in \mathbb{N}$ let $\mathcal{H}^k$ denote the vector subspace of $\mathcal{H}$ which is formed by the k-*times differentiable vectors*

$f \in \mathcal{H}$ for $(U,\mathcal{H})$. Thus $f \in \mathcal{H}$ belongs to $\mathcal{H}^k$ if and only if the mapping $G \in x \rightsquigarrow U(x)f \in \mathcal{H}$ is an element of the space $\mathbb{C}^k(G;\mathcal{H})$. Define

$$\mathcal{H}^\infty = \bigcap_{k \in \mathbb{N}} \mathcal{H}^k .$$

The elements of the vector subspace $\mathcal{H}^\infty$ of $\mathcal{H}$ are called *infinitely differentiable* or *smooth vectors* for $(U,\mathcal{H})$. In this way a descending filtration

$$\mathcal{H}^0 \hookleftarrow \ldots \hookleftarrow \mathcal{H}^k \hookleftarrow \mathcal{H}^{k+1} \hookleftarrow \ldots \hookleftarrow \mathcal{H}^\infty$$

of vector subspaces of $\mathcal{H}^0 = \mathcal{H}$ arises. It is possible to equip $\mathcal{H}^k (k \in \mathbb{N})$ with the structure of a complex Hilbert space such that $\mathcal{H}^k$ becomes isometric to $\mathcal{H}^0$. If $\mathcal{H}^\infty$ is endowed with the projective limit of the topologies of the complex Hilbert spaces $\mathcal{H}^k (k \in \mathbb{N})$, the linear injections of the filtration above are continuous and $\mathcal{H}^\infty$ is everywhere dense in any Hilbert space $\mathcal{H}^k$.

1.3 Let $(U_1,\mathcal{H}_1)$ and $(U_2,\mathcal{H}_2)$ denote any two linear representations of the same group G in the complex vector spaces $\mathcal{H}_1$ and $\mathcal{H}_2$, respectively. A $\mathbb{C}$-linear mapping $T : \mathcal{H}_1 \to \mathcal{H}_2$ is called to be an *intertwining* linear operator or a G-*morphism* for $(U_1,\mathcal{H}_1)$ and $(U_2,\mathcal{H}_2)$ if the identity

$$T \circ U_1(x) = U_2(x) \circ T$$

holds for all $x \in G$. Thus the following diagram

$$\begin{array}{ccc} \mathcal{H}_1 & \xrightarrow{\;T\;} & \mathcal{H}_2 \\ {\scriptstyle U_1(x)}\downarrow & & \downarrow{\scriptstyle U_2(x)} \\ \mathcal{H}_1 & \xrightarrow[\;T\;]{} & \mathcal{H}_2 \end{array}$$

is commutative for *all* $x \in G$. The two linear representations $(U_1,\mathcal{H}_1)$ and $(U_2,\mathcal{H}_2)$ of G are called to be *isomorphic*, if there exists an intertwining linear isomorphism T of $\mathcal{H}_1$ onto $\mathcal{H}_2$, i.e., if the representation spaces $\mathcal{H}_1$ and $\mathcal{H}_2$ are G-isomorphic left G-modules.

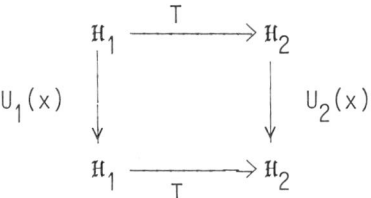

In the case when G is a topological group and $(U_1,\mathcal{H}_1)$, $(U_2,\mathcal{H}_2)$ are two continuous linear representations of G in the complex topological vector spaces $\mathcal{H}_1$ and $\mathcal{H}_2$, respectively, the complex vector space of *continuous*, intertwining, linear operators for $(U_1,\mathcal{H}_1)$ and $(U_2,\mathcal{H}_2)$ is denoted by $\mathcal{R}_G(U_1,U_2)$. In the present case, an *isomorphism* of $(U_1,\mathcal{H}_1)$ onto $(U_2,\mathcal{H}_2)$ is a bicontinuous linear G-isomorphism of $\mathcal{H}_1$ onto $\mathcal{H}_2$. In the special case $U_1 = U_2 = U$, $\mathcal{H}_1 = \mathcal{H}_2 = \mathcal{H}$ obviously $\mathcal{R}_G(U) := \mathcal{R}_G(U,U)$ is the *centralizer* in the complex algebra $\mathrm{End}(\mathcal{H})$ of the image group $U(G)$ of G under U. The invertible elements of the complex algebra $\mathcal{R}_G(U)$ are the *automorphisms* of $(U,\mathcal{H})$.

For two continuous, unitary, linear representations $(U_1,\mathcal{H}_1)$ and $(U_2,\mathcal{H}_2)$ of the topological group G we get by taking Hilbert space adjoints

$$\mathcal{R}_G(U_2,U_1)^* = \mathcal{R}_G(U_1,U_2).$$

In particular, $\mathcal{R}_G(U_1)$ is a weakly closed *subalgebra* of $\mathrm{End}(\mathcal{H}_1)$ that is stable under the involution $T \rightsquigarrow T^*$, hence a von Neumann algebra.

Finally two continuous, unitary, linear representations $(U_1,\mathcal{H}_1)$ and $(U_2,\mathcal{H}_2)$ of the topological group G are said to be *unitarily* isomorphic if there exists a *unitary* linear G-isomorphism S of the complex Hilbert space $\mathcal{H}_1$ onto the complex Hilbert space $\mathcal{H}_2$. It is not difficult to see that any isomorphism T of $(U_1,\mathcal{H}_1)$ onto $(U_2,\mathcal{H}_2)$ gives rise to a *unitary* isomorphism S of $(U_1,\mathcal{H}_1)$ onto $(U_2,\mathcal{H}_2)$. Indeed, $T^* \circ T \in \mathcal{R}_G(U_1)$ is a bicontinuous, Hermitean, positive, linear operator. Hence there exists a unique bicontinuous, Hermitean, positive, linear operator $|T| \in \mathcal{R}_G(U_1)$ such that

$$|T|^2 = T^* \circ T$$

holds. Then $S = T \circ |T|^{-1}$ does the job, to wit, $S \in \mathcal{R}_G(U_1,U_2)$ is a unitary linear isomorphism of $(U_1,\mathcal{H}_1)$ onto $(U_2,\mathcal{H}_2)$ where $S \circ |T|$ forms the *polar decomposition* of T.

1.4 Let $(U,\mathcal{H})$ denote a continuous linear representation of the topological group G in the Hausdorff topological vector space $\mathcal{H}$ over the field $\mathbb{C}$. A vector subspace $\mathcal{L}$ of $\mathcal{H}$ is said to be *stable under* U if $f \in \mathcal{L}$ implies $U(x)f \in \mathcal{L}$ for all elements $x \in G$. In this case $U(G)\mathcal{L} \subseteq \mathcal{L}$, i.e., $\mathcal{L}$ is a G-submodule of $\mathcal{H}$. Of course, the trivial vector subspace $\{0\}$ of $\mathcal{H}$ and the whole vector space $\mathcal{H}$ are always stable under U. If these two vector spaces are the only *closed* vector subspaces of $\mathcal{H}$ stable under U then $(U,\mathcal{H})$ is said to be a *topologically irreducible*, continuous, linear representation of G in $\mathcal{H}$. In this case the topological left G-module $\mathcal{H}$ is topologically simple and the orbits

$$U(G)f \qquad (f \in \mathcal{H} - \{0\})$$

of the linear left G-action on $\mathcal{H}$ associated with $(U,\mathcal{H})$ are *total* subsets of $\mathcal{H}$. If $\mathcal{H}$ is finite dimensional over $\mathbb{C}$, then $\dim_{\mathbb{C}}\mathcal{H}$ is called to be the *dimension* of $(U,\mathcal{H})$. Every vector subspace of $\mathcal{H}$ is closed and therefore $(U,\mathcal{H})$ is simply called an *irreducible*, continuous, linear representation of G in $\mathcal{H}$ in this case.

The set $\hat{G}$ of isomorphy classes of topologically irreducible, continuous, unitary, linear representations of G is called the *unitary dual* or the *Mackey dual* of G. Using this notion, the first main problem of harmonic analysis can be described as follows: Given a locally compact topological group G, determine its unitary dual $\hat{G}$.

Suppose $(U,\mathcal{H})$ is a continuous, linear representation of G in $\mathcal{H}$ that is *not* topologically irreducible. The restrictions of the linear operators $\{U(x) \in \underline{\underline{GL}}(\mathcal{H}) | x \in G\}$ to a *closed* vector subspace $\mathcal{L}$ of $\mathcal{H}$ that is stable under U and satisfies $\{0\} \subsetneqq \mathcal{L} \subsetneqq \mathcal{H}$ define a continuous linear representation $(U,\mathcal{L})$ of G in $\mathcal{L}$ when $\mathcal{L}$ carries the vector space topology induced by $\mathcal{H}$. In this case, $(U,\mathcal{L})$ is said to form a linear subrepresentation $U|\mathcal{L}$ of the given continuous, linear representation $(U,\mathcal{H})$ of G. If $\mathcal{H}$ is the Hilbert sum of a family $(\mathcal{H}_j)_{j\in I}$ of (mutually orthogonal) closed vector subspaces of $\mathcal{H}$ that are stable under U and $((U_j,\mathcal{H}_j))_{j\in I}$ denotes the corresponding family of linear subrepresentations of $(U,\mathcal{H})$, then $(U,\mathcal{H})$ is called to be the *Hilbert sum* of the family $((U_j,\mathcal{H}_j))_{j\in I}$. In this case we shall write

$$U = \hat{\bigoplus_{j \in I}} U_j, \quad \mathcal{H} = \hat{\bigoplus_{j \in I}} \mathcal{H}_j.$$

In particular, $(U,\mathcal{H})$ is unitary if and only if all the linear subrepresentations $((U_j,\mathcal{H}_j))_{j\in I}$ of $(U,\mathcal{H})$ are unitary. If moreover the unitary linear subrepresentations $((U_j,\mathcal{H}_j))_{j\in I}$ are topologically irreducible, then $(U,\mathcal{H})$ is said to admit a *discrete decomposition*. In the case when all the linear subrepresentations $((U_j,\mathcal{H}_j))_{j\in I}$ are isomorphic to the same topologically irreducible, continuous, unitary, linear representation $(U_0,\mathcal{H}_0)$ of G, we use the notation

$$U = |I|.U_0$$

where $|I| = \mathrm{Card}(I)$ denotes the cardinality of the index family I. In this case, $(U, \mathcal{H})$ is said to be a *multiple* of $(U_0, \mathcal{H}_0)$. Any continuous linear representation of G which is isomorphic to a linear subrepresentation of $(U, \mathcal{H})$ is said to be *contained* in $(U, \mathcal{H})$.

The second main problem of harmonic analysis now can be described in the following way: Given a continuous, unitary, linear representation $(U, \mathcal{H})$ of a locally compact topological group G, determine all those isomorphy classes in the unitary dual $\hat{G}$ such that one (and hence every) representative of these classes is contained in $(U, \mathcal{H})$.

1.5 Lemma. Let $(U, \mathcal{H})$ be a continuous linear representation of the topological group G. Let $\mathcal{L}$ denote a closed vector subspace of the complex Hilbert space $\mathcal{H}$. Then $\mathcal{L}$ is stable under U if and only if the orthoprojector $P_{\mathcal{L}}$ of $\mathcal{H}$ onto $\mathcal{L}$ belongs to $\mathcal{R}_G(U)$.

Proof. Suppose that $\mathcal{L}$ is stable under U. For $f \in \mathcal{H}$ we have $P_{\mathcal{L}}(f) \in \mathcal{L}$ and therefore $U(x) \circ P_{\mathcal{L}}(f) \in \mathcal{L}$ for all $x \in G$. It follows $P_{\mathcal{L}} \circ U(x) \circ P_{\mathcal{L}}(f) = U(x) \circ P_{\mathcal{L}}(f)$ for all $f \in \mathcal{H}$ and therefore

$$P_{\mathcal{L}} \circ U(x) \circ P_{\mathcal{L}} = U(x) \circ P_{\mathcal{L}}$$

for all $x \in G$. Since $P_{\mathcal{L}} \in \mathrm{End}(\mathcal{H})$ is a Hermitean endomorphism it follows by taking Hilbert space adjoints

$$P_{\mathcal{L}} \circ U(x) \circ P_{\mathcal{L}} = P_{\mathcal{L}} \circ U(x)$$

for all $x \in G$. Thus $P_{\mathcal{L}} \in \mathcal{R}_G(U)$.

Conversely, the identity $P_{\mathcal{L}} \circ U(x)f = U(x)f$ for all vectors $f \in \mathcal{L}$ shows that $U(x)f \in \mathcal{L}$ holds for all $x \in G$. Thus $\mathcal{L}$ is stable under U. —

Corollary. The closed vector subspace $\mathcal{L}$ of $\mathcal{H}$ is stable under U if and only if its orthogonal complement $\mathcal{L}^{\perp}$ in $\mathcal{H}$ is stable under U.

1.6 Theorem (Schur). Let $(U_1, \mathcal{H}_1)$ and $(U_2, \mathcal{H}_2)$ denote two continuous, unitary, linear representations of the topological group G. Suppose that $(U_1, \mathcal{H}_1)$ is topologically irreducible. Any linear operator $T \in \mathcal{R}_G(U_1, U_2)$ is an isomorphism of $\mathcal{H}_1$ onto $\mathcal{H}_2$ or satisfies $T = 0$.

Proof. In view of the identity $\mathcal{R}_G(U_2, U_1) = \mathcal{R}_G(U_1, U_2)^*$ we conclude that

$T^* \circ T \in \mathfrak{R}_G(U_1)$ is a continuous, Hermitean, positive endomorphism of the complex Hilbert space $\mathcal{H}_1$. An application of the spectral theorem to $T^* \circ T$ yields a one-parameter family $(P_t)_{t\in\mathbb{R}}$ of orthoprojectors in $\mathcal{H}_1$ belonging to $\mathfrak{R}_G(U_1)$. By virtue of Lemma 1.5 we have $P_t = 0$ or $= \mathrm{id}_{\mathcal{H}_1}$ for all $t \in \mathbb{R}$. The spectral theorem shows the existence of a number $t_0 \in \mathbb{R}_+$ such that

$$T^* \circ T = t_0 \mathrm{id}_{\mathcal{H}_1}$$

holds. From this identity the statement becomes obvious.

Corollary 1 (Schur-Naimark). A continuous, unitary, linear representation $(U,\mathcal{H})$ of the topological group G is topologically irreducible if and only if the algebra $\mathfrak{R}_G(U)$ of continuous intertwining operators of $(U,\mathcal{H})$ with itself consists only of the homotheties of $\mathcal{H}$, i.e., if we have

$$\mathfrak{R}_G(U) = \{\zeta.\mathrm{id}_{\mathcal{H}} | \zeta \in \mathbb{C}\}$$

or equivalently

$$\dim_{\mathbb{C}} \mathfrak{R}_G(U) = 1.$$

Proof. Since the spectrum of any continuous endomorphism of $\mathcal{H}$ is a nonempty subset of $\mathbb{C}$, there exists for any $T \in \mathfrak{R}_G(U)$ a number $\zeta \in \mathbb{C}$ such that $T - \zeta.\mathrm{id}_{\mathcal{H}} \in \mathfrak{R}_G(U)$ does not admit an inverse in $\mathrm{End}(\mathcal{H})$. If $(U,\mathcal{H})$ is topologically irreducible, Theorem 1.6 implies $T - \zeta.\mathrm{id}_{\mathcal{H}} = 0$. The converse follows from Lemma 1.5. —

Let G and $(U,\mathcal{H})$ be as in Corollary 1 above. The *projective kernel* $\mathfrak{P}_G(U)$ of $(U,\mathcal{H})$ in G is the set of all elements $x \in G$ such that $U(x)$ is a homothety of $\mathcal{H}$. Let C denote the *centre* of G. Since $U(z) \in \mathfrak{R}_G(U) \cap \underline{\underline{U}}(\mathcal{H})$ for all $z \in C$, the preceding Corollary 1 implies the existence of a continuous morphism $\chi_U : C \to \mathbb{T}$ such that

$$U(z) = \chi_U(z).\mathrm{id}_{\mathcal{H}}$$

holds for all $z \in C$. Hence $C \subseteq \mathfrak{P}_G(U)$. The continuous unitary character χ_U of C is said to be the *central character* of the topologically irreducible, continuous, unitary, linear representation $(U,\mathcal{H})$ of G. In the case when G is an abelian group, i.e., $G = C$, and moreover G is a locally compact topological group, the Mackey dual $\hat{G}$ reduces to the Pontrjagin dual of G.

Corollary 2. Let $(U,\mathcal{H})$ denote a continuous, unitary, linear representation of the topological group G - then $(U,\mathcal{H})$ is topologically irreducible if and only if the centralizer in $\underline{\underline{U}}(\mathcal{H})$ of the image group U(G) of G under U satisfies

$$\mathbb{R}_G(U) \cap \underline{\underline{U}}(\mathcal{H}) = \mathbb{T}.\mathrm{id}_{\mathcal{H}}.$$

Thus the group of unitary automorphisms of $(U,\mathcal{H})$ can be identified with the one-dimensional compact torus group $\mathbb{T}$ if and only if $(U,\mathcal{H})$ is topologically irreducible.

1.7 A continuous, unitary, linear representation $(U,\mathcal{H})$ of the topological group G is said to be *primary* if the *centre* of the complex algebra $\mathbb{R}_G(U)$ of continuous, intertwining, linear operators of $(U,\mathcal{H})$ with itself consists only of the *homotheties* of $\mathcal{H}$. An application of the spectral theorem shows that $(U,\mathcal{H})$ is primary if and only if the centre of $\mathbb{R}_G(U)$ contains no continuous orthoprojector in $\mathcal{H}$ other than the trivial projectors $\{0_{\mathcal{H}}, \mathrm{id}_{\mathcal{H}}\}$. Of course, the topologically irreducible, continuous, unitary, linear representations of G are primary.

A primary representation $(U,\mathcal{H})$ of G is said to be *isotypic* if $(U,\mathcal{H})$ contains a *topologically irreducible* linear subrepresentation of G that is different from the trivial representation $G \ni x \rightsquigarrow \mathrm{id}_{\mathcal{H}}$ of G in $\mathcal{H}$. In order to analyze the isotypic representations of G we shall need the following

1.8 Lemma. Let $(U_1,\mathcal{H}_1)$ and $(U_2,\mathcal{H}_2)$ denote two continuous, unitary, linear representations of the topological group G. Suppose that $(U_1,\mathcal{H}_1)$ is topologically irreducible and $\mathbb{R}_G(U_1,U_2) \neq \{0\}$. Then $(U_2,\mathcal{H}_2)$ contains $(U_1,\mathcal{H}_1)$.

Proof. Let $T \in \mathbb{R}_G(U_1,U_2) - \{0\}$. Then $T^* \circ T \in \mathbb{R}_G(U_1) - \{0\}$ and by virtue of Theorem 1.6 supra we have $T^* \circ T = t_0 \mathrm{id}_{\mathcal{H}_1}$ for a suitable number $t_0 > 0$. It follows

$$\langle Tf|Tg\rangle = \langle T^* \circ Tf|g\rangle = t_0\langle f|g\rangle$$

for all pairs $(f,g) \in \mathcal{H}_1 \times \mathcal{H}_1$. Consequently $T : \mathcal{H}_1 \to \mathcal{H}_2$ is a continuous linear mapping with *closed* range $\mathcal{L}_2 = \mathrm{Im}(T)$ in $\mathcal{H}_2$. Since Ker(T) is stable under U_1 it follows $\mathrm{Ker}(T) = \{0\}$. Therefore T defines a bicontinuous linear isomorphism of $\mathcal{H}_1$ onto the closed vector subspace $\mathcal{L}_2$ of $\mathcal{H}_2$. Since $\mathcal{L}_2$ is stable under U_2, the mapping T defines an isomorphism of $(U_1,\mathcal{H}_1)$ onto the linear subrepresentation $U_2|\mathcal{L}_2$ of $(U_2,\mathcal{H}_2)$. —

More general we have

1.9 Theorem (Topological Version of Schur's Theorem). Let $(U_1,\mathcal{H}_1)$ and $(U_2,\mathcal{H}_2)$ denote two continuous, unitary, linear representations of the topological group G. Suppose that $(U_1,\mathcal{H}_1)$ is topologically irreducible and that there exists a nonzero not necessarily bounded closed linear operator T from $\mathcal{H}_1$ to $\mathcal{H}_2$ such that its domain $\mathcal{D}om(T)$ is everywhere dense in $\mathcal{H}_1$ and stable under U_1, and such that the intertwining property

$$T \circ (U_1(x)f_1) = U_2(x) \circ (Tf_1)$$

holds for all $x \in G$ and $f_1 \in \mathcal{D}om(T)$. Then $\mathcal{D}om(T) = \mathcal{H}_1$, and T has an extension belonging to $\mathcal{R}_G(U_1,U_2)$ such that ξT for a suitable scalar $\xi > 0$ defines a linear G-isometry of $\mathcal{H}_1$ onto a closed vector subspace of $\mathcal{H}_2$. In particular, $(U_2,\mathcal{H}_2)$ contains $(U_1,\mathcal{H}_1)$.

Proof. The vector subspace $\mathcal{D}om(T^*)$ of $\mathcal{H}_2$ is everywhere dense in $\mathcal{H}_2$ and stable under U_2. Similarly $\mathcal{D}om(T^* \circ T)$ is an everywhere dense vector subspace of $\mathcal{H}_1$ and stable under U_1. The spectral theorem shows that the closed self-adjoint positive linear operator $T^* \circ T$ is a homothety on $\mathcal{D}om(T^* \circ T)$. Consequently there exists a real number $t_0 > 0$ such that

$$\langle Tf|Tg\rangle = t_0\langle f|g\rangle$$

holds for all vectors f,g in $\mathcal{D}om(T^* \circ T)$. We conclude that T admits an extension which is a continuous linear mapping of $\mathcal{H}_1$ into $\mathcal{H}_2$ and belongs to $\mathcal{R}_G(U_1,U_2)$. Now the theorem follows easily from the reasoning in Section 1.8.—

1.10 Theorem. Let $(U,\mathcal{H})$ denote a primary representation of the topological group G. The following properties are equivalent:

(i) $(U,\mathcal{H})$ is an isotypic representation of G;

(ii) $U = n.U_0$ where $(U_0,\mathcal{H}_0)$ is a topologically irreducible linear subrepresentation of $(U,\mathcal{H})$ and $n \geqq 1$ is a natural number or $n = +\infty$.

If $(U,\mathcal{H})$ satisfies (ii), it determines the representation $(U_0,\mathcal{H}_0)$ of G uniquely up to unitary isomorphy. Moreover, it determines uniquely the number n. Furthermore, every linear subrepresentation of $(U,\mathcal{H})$ is a multiple of $(U_0,\mathcal{H}_0)$.

Proof. First suppose that $(U,\mathcal{H})$ is an isotypic representation of G. Let $\mathcal{S}_U$ denote the set of all closed vector subspaces of $\mathcal{H}$ that are different from the trivial vector subspace $\{0\}$ and stable under U. Choose an element $\mathcal{H}_0 \in \mathcal{S}_U$ that is minimal with respect to the ordering of $\mathcal{S}_U$ by set theoretic inclusion. We may suppose that $\mathcal{H}_0 \subsetneqq \mathcal{H}$ because otherwise $(U,\mathcal{H})$ would be topologically irreducible and in this case (i) implies (ii) trivially. By virtue of Lemma 1.5 and its Corollary the orthoprojectors $P_{\mathcal{H}_0}$ and $P_{\mathcal{H}_0^\perp} = id_{\mathcal{H}} - P_{\mathcal{H}_0}$ of $\mathcal{H}$ onto $\mathcal{H}_0$ and $\mathcal{H}_0^\perp$, respectively, both belong to $\mathcal{R}_G(U)$, but they do *not* belong to the centre of $\mathcal{R}_G(U)$ since $(U,\mathcal{H})$ is a primary representation of G. It follows that there exists an operator $T \in \mathcal{R}_G(U)$ such that $T(\mathcal{H}_0) \not\subset \mathcal{H}_0$ holds. Consequently we have

$$P_{\mathcal{H}_0^\perp} \circ T \in \mathcal{R}_G(U|\mathcal{H}_0, U|\mathcal{H}_0^\perp) - \{0\}.$$

By virtue of Lemma 1.8 the linear subrepresentation $U|\mathcal{H}_0^\perp$ of $(U,\mathcal{H})$ contains the topologically irreducible linear subrepresentation $U_0 = U|\mathcal{H}_0$. Moreover, if $\mathcal{L} \in \mathcal{S}_U$ is different from $\mathcal{H}$ and the linear subrepresentation $U|\mathcal{L}$ of $(U,\mathcal{H})$ contains a multiple of $(U_0,\mathcal{H}_0)$ then $U|\mathcal{L}^\perp$ contains $(U_0,\mathcal{H}_0)$. An application of Zorn's theorem establishes that $U = n.U_0$ holds. The natural number $n \geqq 1$ is finite if and only if $\dim_{\mathbb{C}} \mathcal{R}_G(U) < +\infty$ and in this case we have

$$\dim_{\mathbb{C}} \mathcal{R}_G(U) = n^2.$$

Thus $(U,\mathcal{H})$ determines the number n uniquely.

Let $\mathfrak{M} \in \mathcal{S}_U$ be arbitrary. There exists a topologically irreducible linear subrepresentation $(V,\mathcal{N})$ of $(U,\mathcal{H})$ which is isomorphic to $(U_0,\mathcal{H}_0)$ such that $P_{\mathfrak{M}}(\mathcal{N}) \neq \{0\}$. Thus the restriction $P_{\mathfrak{M}}|\mathcal{N}$ belongs to $\mathcal{R}_G(U|\mathcal{N}, U|\mathfrak{M})$. By virtue of Lemma 1.8 the linear subrepresentation $U|\mathfrak{M}$ contains $(U_0,\mathcal{H}_0)$. Therefore $U|\mathfrak{M} = m.U_0$ where $m \geqq 1$ denotes a natural number or $m = +\infty$. In particular, every topologically irreducible linear subrepresentation of $(U,\mathcal{H})$ is isomorphic to $(U_0,\mathcal{H}_0)$. Thus $(U,\mathcal{H})$ determines $(U_0,\mathcal{H}_0)$ uniquely up to unitary isomorphy.

Now suppose that $(U_0,\mathcal{H}_0)$ is a topologically irreducible linear subrepresentation of $(U,\mathcal{H})$ and $U = n.U_0$. Let I denote an index family of cardinality $|I| = n$ and $(\mathcal{H}_j)_{j\in I}$ a family of closed vector subspaces of $\mathcal{H}$ stable under U such that $\mathcal{H} = \hat{\bigoplus}_{j \in I} \mathcal{H}_j$ holds and $(U_j,\mathcal{H}_j)$ where $U_j = U|\mathcal{H}_j$ is isomorphic to $(U_0,\mathcal{H}_0)$ for all $j \in I$. The orthoprojectors $P_{\mathcal{H}_j}$ of $\mathcal{H}$ onto $\mathcal{H}_j$ belong to $\mathcal{R}_G(U)$ for all $j \in I$ by Lemma 1.5. Let $P \neq 0_{\mathcal{H}}$ be an arbitrary orthoprojector in $\mathcal{H}$ that belongs to the centre of $\mathcal{R}_G(U)$. Then $P \circ P_{\mathcal{H}_j}$ and $(id_{\mathcal{H}} - P) \circ P_{\mathcal{H}_j}$ are orthoprojectors in $\mathcal{H}$ that belong to $\mathcal{R}_G(U_j)$ for all $j \in I$. By virtue of the topological irreducibility of $(U_j,\mathcal{H}_j)$ we have either $P \circ P_{\mathcal{H}_j} = 0_{\mathcal{H}}$ or $P \circ P_{\mathcal{H}_j} = P_{\mathcal{H}_j}$ for $j \in I$. Hence there exists at least one index $j_0 \in I$ such that $P \circ P_{\mathcal{H}_{j_0}} = P_{\mathcal{H}_{j_0}}$ holds true since otherwise we would have $P = 0_{\mathcal{H}}$. For any $k \in I - \{j_0\}$ we have

$$\{0\} \subsetneqq \mathcal{R}_G(U_{j_0},U_k) \subseteqq \mathcal{R}_G(U)$$

since $(U_{j_0},\mathcal{H}_{j_0})$ and $(U_k,\mathcal{H}_k)$ are isomorphic linear subrepresentations of $(U,\mathcal{H})$. Let $T \in \mathcal{R}_G(U_{j_0},U_k) - \{0\}$ be arbitrary. Then we have

$$P \circ P_{\mathcal{H}_k} \circ T \circ P_{\mathcal{H}_{j_0}} = P \circ T \circ P_{\mathcal{H}_{j_0}} = T \circ P \circ P_{\mathcal{H}_{j_0}} = T \circ P_{\mathcal{H}_{j_0}} \neq 0_{\mathcal{H}}$$

and therefore $P \circ P_{\mathcal{H}_k} \neq 0_{\mathcal{H}}$ for all $k \in I - \{j_0\}$. Consequently, the identity $P \circ P_{\mathcal{H}_j} = P_{\mathcal{H}_j}$ holds for all $j \in I$, i.e., we have $P = id_{\mathcal{H}}$. Thus $(U,\mathcal{H})$ is primary and since it contains $(U_0,\mathcal{H}_0)$ it forms an isotypic representation of G. —

In the following we shall call the natural numbers $n \geqq 1$ or $n = +\infty$ in Theorem 1.10 the *multiplicity* of $(U_0,\mathcal{H}_0)$ in $(U,\mathcal{H})$.

Remark. In Section 7 infra we shall see that the representations of the real Heisenberg nilpotent Lie group $\tilde{A}(\mathbb{R}^n)$ unitarily induced by its central characters are isotypic.

1.11 Let $(U,\mathcal{H})$ be a continuous linear representation of the topological group G. A vector $f_0 \in \mathcal{H}$ is said to be *cyclic* or a *totalizer* for $(U,\mathcal{H})$ if the orbit $U(G)f_0$ of f_0 for the canonical G-action on $\mathcal{H}$ is total in $\mathcal{H}$. A continuous linear representation $(U,\mathcal{H})$ of G is said to be *topologically cyclic* (or *monogenic*) if it has a cyclic vector $f_0 \in \mathcal{H}$. Of course, $(U,\mathcal{H})$ is topologically irreducible if and only if all the vectors $f \in \mathcal{H} - \{0\}$ are totalizers for $(U,\mathcal{H})$.

A continuous, unitary, linear representation $(U,\mathcal{H})$ of a topological group does not necessarily admit a discrete decomposition in the sense of Section 1.4. supra. However we have

1.12 Lemma. Any continuous, unitary, linear representation $(U,\mathcal{H})$ of the topological group G is a Hilbert sum of a family of topologically cyclic, continuous, unitary, linear representations of G.

Proof. Let $\mathfrak{S}_U$ denote the set of all vector subspaces $\neq \{0\}$ of $\mathcal{H}$ that are Hilbert sums of closed vector subspaces of $\mathcal{H}$ such that each vector subspace is stable under U and admits a cyclic vector for the restriction of U. Then the ordering of $\mathfrak{S}_U$ by set theoretic inclusion is inductive. An application of Zorn's theorem furnishes a maximal element in $\mathfrak{S}_U$. It is immediate that this maximal element coincides with $\mathcal{H}$. ——

1.13 Lemma. Let $(U_\alpha,\mathcal{H}_\alpha)$, $\alpha \in \{1,2\}$, denote two continuous, unitary, linear representations of the topological group G. Suppose that there exists a cyclic vector $f_\alpha \in \mathcal{H}_\alpha$ for $(U_\alpha,\mathcal{H}_\alpha)$ such that the identity

$$\langle U_1(x)f_1|f_1\rangle = \langle U_2(x)f_2|f_2\rangle$$

holds for all elements $x \in G$. Then there exists a unitary linear isomorphism S of $(U_1,\mathcal{H}_1)$ onto $(U_2,\mathcal{H}_2)$ such that $S(f_1) = f_2$.

Proof. For any finite subset $\{x_j \in G \mid j \in I\}$ of points of G and any set of scalars $\{\lambda_j \in \mathbb{C} | j \in I\}$ we have

$$\begin{aligned}\|\sum_{j\in I} \lambda_j U_1(x_j)f_1\|^2 &= \sum_{(j,k)\in I\times I} \bar{\lambda}_k\lambda_j\langle U_1(x_k^{-1}x_j)f_1|f_1\rangle \\ &= \sum_{(j,k)\in I\times I} \bar{\lambda}_k\lambda_j\langle U_2(x_k^{-1}x_j)f_2|f_2\rangle \\ &= \|\sum_{j\in I} \lambda_j U_2(x_j)f_2\|^2 .\end{aligned}$$

In particular, $\sum_{j\in I} \lambda_j U_1(x_j)f_1 = 0$ is equivalent to $\sum_{j\in I} \lambda_j U_2(x_j)f_2 = 0$. It follows that the $\mathbb{C}$-linear mapping S_0 of the vector subspace of $\mathcal{H}_1$ spanned by the orbit $U_1(G)f_1$ of $f_1 \in \mathcal{H}_1$ onto the vector subspace of $\mathcal{H}_2$ spanned by the orbit $U_2(G)f_2$ of $f_2 \in \mathcal{H}_2$ is well defined by the prescription

$$S_0\left(\sum_{j\in I} \lambda_j U_1(x_j)f_1\right) = \sum_{j\in I} \lambda_j U_2(x_j)f_2$$

and maps the cyclic vectors $f_\alpha \in \mathcal{H}_\alpha$ correctly, i.e.,

$$S_0(f_1) = f_2.$$

Since S_0 is an isometric linear mapping, i.e., a $\mathbb{C}$-linear G-isomorphism between complex prehilbert spaces that are everywhere dense in $\mathcal{H}_1$ and $\mathcal{H}_2$, respectively, it admits a unique linear extension $S \in \mathcal{R}_G(U_1,U_2)$ that maps the Hilbert space $\mathcal{H}_1$ onto the Hilbert space $\mathcal{H}_2$. —

1.14 Let $(U,\mathcal{H})$ denote a continuous, unitary, linear representation of the locally compact topological group G. For any pair $(f,g) \in \mathcal{H} \times \mathcal{H}$ the function

$$c_{U,f,g} : G \ni x \rightsquigarrow \langle U(x)f|g\rangle \in \mathbb{C}$$

is called the *coefficient of* $(U,\mathcal{H})$ *with respect to the pair* $(f,g) \in \mathcal{H} \times \mathcal{H}$. Clearly, $c_{U,f,g}$ is a continuous complex-valued function on G that is *bounded* by $\|f\| \cdot \|g\|$.

Let $\mathfrak{M}^1(G)$ denote the complex vector space of bounded (Radon) measures on G. Thus $\mathfrak{M}^1(G)$ is the topological dual $\mathfrak{C}_0'(G)$ of the complex Banach space $\mathfrak{C}_0(G)$ of all continuous complex-valued functions on G vanishing at infinity. Under the convolution product

$$(\mu,\nu) \rightsquigarrow \mu * \nu$$

where $\langle\phi,\mu*\nu\rangle = \iint_{G\times G} \phi(xy)d\mu(x)d\nu(x)$ for $\phi \in \mathfrak{C}_0(G)$, and the 'total mass' norm

$$\mu \rightsquigarrow \|\mu\|_1 = \int_G d|\mu|(x) = |\mu|(G)$$

the vector space $\mathfrak{M}^1(G)$ forms a complex Banach algebra with unit element ε_{1_G} (Dirac measure concentrated at the neutral element 1_G of G). For any measure $\mu \in \mathfrak{M}^1(G)$ the sesquilinear form

$$\mathcal{H} \times \mathcal{H} \ni (f,g) \rightsquigarrow \int_G c_{U,f,g}(x)d\mu(x) \in \mathbb{C}$$

is continuous and of norm $\leq \|\mu\|_1$. It follows that there exists a unique continuous linear operator $U^1(\mu) \in \mathrm{End}(\mathcal{H})$ of norm $\|U^1(\mu)\| \leq \|\mu\|_1$ such that

$$\langle U^1(\mu)f|g\rangle = \int_G c_{U,f,g}(x)d\mu(x)$$

holds for all pairs $(f,g) \in \mathcal{H} \times \mathcal{H}$. Thus $U^1(\mu)f = \int_G (U(x)f)d\mu(x)$ for all $f \in \mathcal{H}$. It is convenient to write this relation in the form

$$U^1(\mu) = \int_G U(x)d\mu(x).$$

We then call $U^1(\mu)$ the *integrated form* of U. Obviously $U^1(\varepsilon_x) = U(x)$ holds for all $x \in G$ and in particular $U(\varepsilon_{1_G}) = \mathrm{id}_{\mathcal{H}}$. Thus U^1 *extends* U from G to $\mathfrak{M}^1(G)$. Moreover we obtain for all measures $\mu \in \mathfrak{M}^1(G)$ and all vectors $f \in \mathcal{H}$ the identities

$$\begin{aligned} U^1(\mu*\nu)f &= \int_G U(x)fd(\mu*\nu)(x) \\ &= \iint_{G\times G} U(xy)fd\mu(x)d\nu(y) \\ &= \iint_{G\times G} U(x) \circ U(y)fd\mu(x)d\nu(y) \\ &= \int_G U^1(\mu) \circ U(y)fd\nu(y) \\ &= U^1(\mu) \circ (\int_G U(y)fd\nu(y)) \\ &= U^1(\mu) \circ U^1(\nu)f. \end{aligned}$$

Thus we established the relation

$$U^1(\mu*\nu) = U^1(\mu) \circ U^1(\nu)$$

for all pairs $(\mu,\nu) \in \mathfrak{M}^1(G) \times \mathfrak{M}^1(G)$. For any measure $\mu \in \mathfrak{M}^1(G)$ let $\check{\mu}$ denote its image under the topological antiautomorphism $x \rightsquigarrow x^{-1}$ of G. Put

$$\mu^* = \bar{\check{\mu}} = \check{\bar{\mu}}.$$

The mapping $\mu \rightsquigarrow \mu^*$ defines a norm preserving involution on $\mathfrak{M}^1(G)$. A straightforward computation shows the identity

$$U^1(\mu^*) = U^1(\mu)^*$$

for all $\mu \in \mathfrak{M}^1(G)$. Thus the mapping $\mu \to U^1(\mu)$ defines a (continuous) morphism of the complex Banach algebra $\mathfrak{M}^1(G)$ into the complex Banach algebra $\mathrm{End}(\mathcal{H})$ which commutes with their involutions and satisfies $U^1(\varepsilon_{1_G}) = \mathrm{id}_{\mathcal{H}}$. Such a morphism is called a *representation* $(U^1,\mathcal{H})$ of the involutory Banach algebra $\mathfrak{M}^1(G)$ over $\mathbb{C}$ in the complex Hilbert space $\mathcal{H}$. The preceding reasoning shows that $(U^1,\mathcal{H})$ extends $(U,\mathcal{H})$ from G to $\mathfrak{M}^1(G)$.

1.15 Theorem (Extension principle). There is a bijection between the set of continuous, unitary, linear representations $(U,\mathcal{H})$ of the locally compact topological group G and the set of representations $(U^1,\mathcal{H})$ of the involutory complex Banach algebra $\mathfrak{M}^1(G)$ in the complex Hilbert space $\mathcal{H}$.

Proof. In view of the preceding reasoning it will be sufficient to show that a representation $(R,\mathcal{H})$ of $\mathfrak{M}^1(G)$ in a complex Hilbert space $\mathcal{H}$ gives rise to a continuous, unitary, linear representation $(U,\mathcal{H})$ of G such that $U^1 = R$ holds. Define $(U,\mathcal{H})$ according to the rule $U : G \ni x \rightsquigarrow R(\varepsilon_x) \in \mathrm{End}(\mathcal{H})$. Then we get $U(1_G) = R(\varepsilon_{1_G}) = \mathrm{id}_{\mathcal{H}}$ and for all pairs $(x,y) \in G \times G$ the relations

$$U(xy) = R(\varepsilon_{(xy)}) = R(\varepsilon_x * \varepsilon_y) = R(\varepsilon_x) \circ R(\varepsilon_y) = U(x) \circ U(y).$$

Moreover, we have

$$U(x^{-1}) = R(\varepsilon_{x^{-1}}) = R(\varepsilon_x^*) = R(\varepsilon_x)^* = U(x)^*$$

for all $x \in G$. Consequently, $(U,\mathcal{H})$ defines a unitary linear representation of G in $\mathcal{H}$, the continuity of which can be easily established. Since $\mu = \int_G (\varepsilon_x * \varepsilon_{1_G}) d\mu(x)$ holds for all $\mu \in \mathfrak{M}^1(G)$, the proof is complete. ——

1.16 Let G be a locally compact topological group and dx a left Haar measure on G. Since the complex vector space $L^1(G;dx)$ does not depend upon the specific choice of the left Haar measure dx of G we denote it by $L^1(G)$

for short. Embed the complex Banach space $L^1(G)$ as a closed left ideal into the complex Banach algebra $\mathfrak{M}^1(G)$ of bounded complex measures on G by means of the isometric morphism

$$\Phi : \phi \rightsquigarrow \phi\cdot dx$$

which associates with every element $\phi \in L^1(G)$ the bounded measure on G with density ϕ and base dx. Let $\Delta_G : G \to \mathbb{R}_+^*$ denote the right *modular function of* G. We remind the reader that Δ_G is a continuous morphism of G into the multiplicative group $\mathbb{R}_+^*$. We then obtain

$$\phi^* = \check{\Delta}_G \, \overset{\vee}{\bar{\phi}}$$

and $\phi \rightsquigarrow \phi^*$ defines an isometric involution of the complex Banach algebra $L^1(G)$. A straightforward calculation shows that the map $\Phi: L^1(G) \to \mathfrak{M}^1(G)$ is a morphism of involutory Banach algebras over $\mathbb{C}$.

For each element $\phi \in L^1(G)$ and any continuous, unitary, linear representation $(U,\mathcal{H})$ of G the integrated form $U^1(\phi) := U^1(\Phi(\phi)) = \int_G \phi(x).U(x)dx$ of U given by

$$U^1(\phi) : \mathcal{H} \ni f \rightsquigarrow \int_G \phi(x).(U(x)f)dx \in \mathcal{H}$$

is linear and continuous. The continuous morphism $\phi \rightsquigarrow U^1(\phi)$ of the complex involutory Banach algebra $L^1(G)$ into the complex operator algebra End$(\mathcal{H})$ is a *nondegenerate* representation $(U^1,\mathcal{H})$ of $L^1(G)$ in $\mathcal{H}$, i.e., $U(L^1(G))f_0 = \{0\}$ for $f_0 \in \mathcal{H}$ implies $f_0 = 0$. Indeed, let $\mathfrak{B}$ denote a filter base of compact neighbourhoods of 1_G in G and $(\phi_V)_{V\in\mathfrak{B}}$ a continuous approximate unit of the complex Banach algebra $L^1(G)$. Thus we suppose that for every neighbourhood $V \in \mathfrak{B}$ the function ϕ_V is an element of the vector subspace $\mathbb{K}(G)$ of $\mathbb{C}_0(G)$ formed by the functions $\phi \in \mathbb{C}_0(G)$ vanishing in a neighbourhood of infinity, i.e., having *compact support* Supp(ϕ) in G. Moreover we suppose Supp$(\phi_V) \subseteq V$ and $\int_G \phi_V(x)dx = 1$ for all $V \in \mathfrak{B}$. The continuity of the unitary linear representation $(U,\mathcal{H})$ combined with the identity

$$\begin{aligned} U^1(\phi_V)f-f &= \int_G \phi_V(x).(U(x)f)dx - \left(\int_G \phi_V(x)dx\right)f \\ &= \int_G \phi_V(x).(U(x)f-f)dx \end{aligned}$$

which is valid for all $f \in \mathcal{H}$ and $V \in \mathcal{B}$ shows that

$$\lim_{V \in \mathcal{B}} \|U^1(\phi_V)f - f\| = 0$$

holds. Thus if $U^1(\phi)f_0 = 0$ for $f_0 \in \mathcal{H}$ and all $\phi \in L^1(G)$, the conclusion $f_0 = 0$ is immediate.

If $\phi \in \mathcal{K}(G)$ then $U^1(\phi)$ is said to be obtained by *smearing* the linear representation $(U,\mathcal{H})$ of G with the continuous, compactly supported, complex-valued function ϕ on G. For any pair $(f,g) \in \mathcal{H} \times \mathcal{H}$ the continuous linear form

$$c_{U^1,f,g} : L^1(G) \ni \phi \rightsquigarrow \langle U^1(\phi)f|g\rangle = \int_G \phi(x)c_{U,f,g}(x)dx \in \mathbb{C}$$

is called the *coefficient of the integrated form* $(U^1,\mathcal{H})$ *of* $(U,\mathcal{H})$ *with respect to the pair* $(f,g) \in \mathcal{H} \times \mathcal{H}$. The functions $c_{U^1,f_0,f_0} \in L^\infty(G)$ $f_0 \in \mathcal{H}$, determine $(U,\mathcal{H})$ up to unitary isomorphy.

In the present case the analog of Theorem 1.15 reads as follows:

1.17 Theorem (Restricted extension principle). There is a bijection between the set of continuous, unitary, linear representations $(U,\mathcal{H})$ of the locally compact topological group G and the set of nondegenerate representations $(U^1,\mathcal{H})$ of the involutory complex Banach algebra $L^1(G)$ in the complex Hilbert space $\mathcal{H}$.

Proof. It remains to prove that, for each nondegenerate representation $(R,\mathcal{H})$ of the involutory complex Banach algebra $L^1(G)$ in the complex Hilbert space $\mathcal{H}$, there exists a unique, continuous, unitary, linear representation $(U,\mathcal{H})$ of G such that $R(\phi) = U^1(\phi)$ holds for all $\phi \in L^1(G)$. Since the notion of topologically cyclic representations of an involutory complex algebra is defined analogously as in the case of continuous linear representations of a topological group (cf. Section 1.11) we may transfer Lemma 1.12 to nondegenerate representations of involutory complex algebras. Thus we may suppose without restricting the generality that $(R,\mathcal{H})$ is a topologically cyclic representation of $L^1(G)$ in $\mathcal{H}$. Choose a totalizer $f_0 \in \mathcal{H}$ of $(R,\mathcal{H})$ and define the everywhere dense vector subspace

$$\mathcal{L}_{f_0} = R(L^1(G))f_0$$

of $\mathcal{H}$. Then the prescription

$$U(x)(R(\phi)f_0) = R(\varepsilon_x * \phi)f_0 \qquad (x \in G)$$

defines a continuous linear representation $x \rightsquigarrow U(x)$ of G in the complex prehilbert space $\mathcal{L}_{f_0}$. Indeed, $(U,\mathcal{L}_{f_0})$ is well defined since an approximate unit argument shows that the identity $R(\phi)f_0 = R(\psi)f_0$ for ϕ,ψ in $L^1(G)$ implies $R(\varepsilon_x * \phi)f_0 = R(\varepsilon_x * \psi)f_0$ for all $x \in G$. Since we have $(\varepsilon_x * \phi)^\vee * (\varepsilon_x * \psi) = \phi^\vee * \psi$ we conclude that U(x) is an isometry of the complex prehilbert space $\mathcal{L}_{f_0}$ for all $x \in G$. Thus $(U,\mathcal{L}_{f_0})$ extends to a continuous, unitary, linear representation $(U,\mathcal{H})$ of G. In the next step we will show that $R(\phi) = U^1(\phi)$ holds for all $\phi \in L^1(G)$. Indeed, for an arbitrary vector $g \in \mathcal{H}$ there exists a function $\rho \in L^\infty(G)$ depending upon $f_0 \in \mathcal{H}$ and $g \in \mathcal{H}$ such that the identity

$$\langle R(\psi)f_0 | g\rangle = \langle \psi,\rho\rangle = \int_G \psi(x)\rho(x)dx$$

holds for all $\psi \in L^1(G)$. An application of the Lebesgue-Fubini theorem then yields

$$\begin{aligned}
\langle R(\phi) \circ R(\psi)f_0 | g\rangle &= \langle R(\phi * \psi)f_0 | g\rangle \\
&= \int_G \phi * \psi(x)\,\rho(x)dx \\
&= \int_G \rho(x) \int_G \phi(y)\psi(y^{-1}x)dydx \\
&= \int_G \phi(y) \int_G (\varepsilon_y * \psi)(x)\,\rho(x)dxdy \\
&= \int_G \phi(y)\,\langle R(\varepsilon_y * \psi)f_0 | g\rangle dy \\
&= \int_G \phi(y)\,\langle U(y)(R(\psi)f_0) | g\rangle dy \\
&= \langle U^1(\phi) \circ R(\psi)f_0 | g\rangle
\end{aligned}$$

for all ϕ and ψ in $L^1(G)$. Hence $R(\phi) = U^1(\phi)$. Finally we will show that $(R,\mathcal{H})$ determines $(U,\mathcal{H})$ uniquely. Indeed if $(\phi_V)_{V\in\mathcal{B}}$ denotes a continuous approximate unit in $L^1(G)$ then

$$\lim_{V\in\mathcal{B}} \|U^1(\varepsilon_x * \phi_V)f - U(x)f\| = 0$$

holds for all $f \in \mathcal{H}$ and $x \in G$. From this we conclude that $(U,\mathcal{H})$ admits a unique extension $(R,\mathcal{H})$ to $L^1(G)$. —

In the next section we shall summarize some useful criteria for the irreducibility of unitary group representations.

1.18 Theorem. Let $(U,\mathcal{H})$ denote a continuous, unitary, linear representation of the locally compact topological group G. The following conditions are mutually equivalent:

(i) $(U,\mathcal{H})$ is topologically irreducible.

(ii) The complex algebra $\mathbb{R}_G(U)$ is one-dimensional, i.e., isomorphic to the field $\mathbb{C}$.

(iii) $U(G)$ is a total subset of $\mathrm{End}(\mathcal{H})$ with respect to the strong operator topology.

(iv) $U^1(\mathbb{K}(G))$ is everywhere dense in $\mathrm{End}(\mathcal{H})$ with respect to the strong operator topology.

Proof. (i) $\Leftrightarrow$ (ii) follows from Corollary 1 of Theorem 1.6. (ii) $\Rightarrow$ (iii) is the von Neumann density theorem. (iii) $\Rightarrow$ (iv) follows from the fact already used in the proof of Theorem 1.17 that for any continuous approximate unit $(\phi_V)_{V\in\mathcal{B}}$ in $L^1(G)$ and for any $x \in G$ the operator $U(x) \in \underline{\underline{U}}(\mathcal{H})$ is the strong limit of $U(\varepsilon_x * \phi_V)$ for $V \in \mathcal{B}$. Finally (iv) $\Rightarrow$ (ii) since the commutant of $\mathrm{End}(\mathcal{H})$ is one-dimensional over $\mathbb{C}$. —

1.19 Let G denote a group. A function $\phi: G \to \mathbb{C}$ is called to be *of positive type on* G if and only if for every *finite* set $\{x_j \in G \mid j \in I\}$ of points of G and every set of scalars $\{\lambda_j \in \mathbb{C} \mid j \in I\}$ we have

$$\sum_{(j,k)\in I\times I} \phi(x_k^{-1}x_j)\bar{\lambda}_k \lambda_j \geq 0.$$

Thus ϕ is of positive type on G if and only if the sesquilinear form on the vector space $\mathbb{C}^{|I|}$ of matrix $(\phi(x_k^{-1}x_j))_{(j,k)\in I\times I}$ with respect to the canonical basis of $\mathbb{C}^{|I|}$ is a positive (semi-definite) Hermitean form for *all* choices of the finite subsets $\{x_j \mid j \in I\}$ of G. Of course, this form may be degenerate. It follows

(I) $\phi(1_G) \geqq 0$

and by considering the matrix

$$\begin{pmatrix} \phi(1_G) & \phi(x) \\ \phi(x^{-1}) & \phi(1_G) \end{pmatrix}$$

associated with the two-point subset $\{1_G, x\}$ of G we obtain

(II) $\overset{\vee}{\bar{\phi}} = \phi$

(central Hermitean symmetry), and

$$||\phi||_\infty = \sup_{x \in G} |\phi(x)| = \phi(1_G)$$

(central peak property).

If G denotes a topological group we denote by $\mathfrak{P}(G)$ the *convex cone in* $\mathfrak{C}(G)$ of all *continuous complex-valued functions of positive type on* G. In the case of a locally compact topological group G, choose a left Haar-measure dx on G. Then the following characterizations hold.

1.20 Theorem. Let G be a locally compact topological group and $\phi \in \mathfrak{C}(G)$. Then the following conditions are mutually equivalent:

(i) $\phi \in \mathfrak{P}(G)$;

(ii) For every function $\psi \in \mathbb{K}(G)$ we have $\langle \phi, \psi^* \star \psi \rangle \geqq 0$;

(iii) The function ϕ is bounded on G and for every measure $\mu \in \mathfrak{M}^1(G)$ we have $\langle \phi, \mu^* \star \mu \rangle \geqq 0$.

Proof. (i) ⇒ (ii): For every function $\psi \in \mathbb{K}(G)$ the function

$$\Psi : G \times G \ni (y,z) \rightsquigarrow \phi(y^{-1}z)\bar{\psi}(y)\psi(z) \in \mathbb{C}$$

belongs to $\mathbb{K}(G \times G)$. Choose a suitable compact subset K of G such that $\mathrm{Supp}(\Psi) \subseteqq K \times K$. Let $\mathfrak{M}^0(G) = (\mathbb{C}^G)'$ be the complex vector space of *finitely supported* measures on G. The measure induced on K × K by the left Haar measure dx ⊗ dx of G × G is vaguely adherent to the measures $\nu \otimes \nu \in \mathfrak{M}^0(G \times G)$

where $\nu \in \mathfrak{M}^o(G)$ belongs to a set of positive and norm-bounded measures on G. If ν admits the form $\sum_{j\in I} \lambda_j \varepsilon_{x_j}$ with masses $\{\lambda_j \in \mathbb{C} | j \in I\}$ placed at the points $\{x_j \in G | j \in I\}$ of Supp(ν), we have

$$\iint_{G\times G} \Psi(y,z)d\nu(y)d\nu(z) = \sum_{(j,k)\in I\times I} \phi(x_k^{-1}x_j)\bar{\psi}(x_k)\bar{\psi}(x_j)\bar{\lambda}_k\lambda_j \geqq 0.$$

Hence $\iint_{G\times G} \phi(y^{-1}z)\bar{\psi}(y)\psi(z)dydz = \langle\phi,\psi^* * \psi\rangle \geqq 0$ for every function $\psi \in \mathbb{K}(G)$.

(ii) ⇒ (iii): Let $\mathfrak{M}^c(G)$ denote the vector subspace of $\mathfrak{M}^1(G)$ formed by the *compactly supported* measures on G. Choose $\nu \in \mathfrak{M}^o(G)$ and a continuous approximate unit $(\phi_V)_{V\in\mathfrak{B}}$ in $L^1(G)$. Then $\nu * \phi_V \in \mathbb{K}(G)$ for all $V \in \mathfrak{B}$. Hence

$$0 \leqq \langle\phi,\phi_V^* * \nu^* * \nu * \phi_V\rangle = \int_G\int_G\int_G\int_G \phi(xyzt)\phi_V^*(x)\phi_V(t)dxdtd\nu^*(y)d\nu(z).$$

Since $\iint_{G\times G} \phi(xyzt)\phi_V^*(x)\phi_V(t)dxdt$ converges to $\phi(yz)$ uniformly on every compact subset of $G \times G$, we have $\langle\phi, \nu * \nu^*\rangle \geqq 0$ for every measure $\nu \in \mathfrak{M}^c(G)$. In particular, if we choose $\nu \in \mathfrak{M}^o(G)$ we see that $\phi \in \mathfrak{P}(G)$ and $\|\phi\|_\infty = \phi(1_G)$ holds. Since every measure $\mu \in \mathfrak{M}^1(G)$ is adherent to $\mathfrak{M}^c(G)$ under the norm topology and ϕ is bounded on G we conclude that $\langle\phi,\mu^* * \mu\rangle$ is adherent to $\{\langle\phi,\nu^* * \nu\rangle | \nu \in \mathfrak{M}^c(G)\}$ and therefore non-negative.

(iii)⇒(i) is trivial in view of the inclusion $\mathfrak{M}^o(G) \subseteq \mathfrak{M}^1(G)$. —

Let $(U,\mathcal{H})$ denote a continuous, unitary, linear representation of the topological group G in the complex Hilbert space $\mathcal{H}$. Then the coefficient function c_{U,f_0,f_0} belongs to $\mathfrak{P}(G)$ for all vectors $f_0 \in \mathcal{H}$. Indeed, in the proof of Lemma 1.13 supra it was shown that

$$\sum_{(j,k)\in I\times I} c_{U,f_0,f_0}(x_k^{-1}x_j)\bar{\lambda}_k\lambda_j = \|\sum_{j\in I} \lambda_j U(x_j)f_0\|^2 \geqq 0$$

holds. Conversely, the following result will be stablished.

1.21 Theorem (Gel'fand-Segal). Let G be a locally compact topological group and let the function $\phi \in \mathfrak{P}(G)$ be given. There exists a continuous, unitary, linear representation $(U_o,\mathcal{H}_o)$ of G and a cyclic vector $f_o \in \mathcal{H}_o$ with respect to $(U_o,\mathcal{H}_o)$ such that the identity

$$\phi = c_{U_o,f_o,f_o}$$

holds.

Proof. Recall that $\mathfrak{m}^0(G)$ forms an involutory subalgebra of $\mathfrak{m}^1(G)$. Define the vector subspace $\mathcal{L}_\phi$ of $\mathfrak{C}(G)$ by

$$\mathcal{L}_\phi = \{\mu \star \phi \mid \mu \in \mathfrak{m}^0(G)\} = \mathfrak{m}^0(G) \star \phi$$

which is spanned by the left translations of ϕ. Then

$$N_\phi = \{\mu \star \phi \in \mathcal{L}_\phi \mid \langle \phi, \mu^* \star \mu \rangle = \int_G \bar{\mu} \star \phi(x) d\mu(x) = 0\}$$

forms a vector subspace of $\mathcal{L}_\phi$. Let $p: \mathcal{L}_\phi \to \mathcal{L}_\phi/N_\phi$ denote the canonical surjection of $\mathcal{L}_\phi$ onto the quotient space $\mathcal{L}_\phi/N_\phi$. For the elements $f = \mu \star \phi$, $g = \nu \star \phi$ of $\mathcal{L}_\phi$ define on $\mathcal{L}_\phi/N_\phi$ the sesquilinear form $\langle \cdot | \cdot \rangle_\phi$ by the prescription

$$\langle p(f) | p(g) \rangle_\phi = \langle \phi, \nu^* \star \mu \rangle.$$

We shall conclude that $\langle \cdot | \cdot \rangle_\phi$ defines a scalar product on $\mathcal{L}_\phi/N_\phi$. Indeed, if we have $\mu = \sum_{j \in I} \lambda_j \varepsilon_{x_j}$ and $\nu = \sum_{k \in K} \rho_k \varepsilon_{y_k}$ in $\mathfrak{m}^0(G)$ then

$$\langle p(f) | p(g) \rangle = \int_G \bar{\nu} \star \phi(x) d\mu(x) = \sum_{(j,k) \in I \times K} \phi(y_k^{-1} x_j) \bar{\rho}_k \lambda_j$$

and N_ϕ forms the null space of $\langle \cdot | \cdot \rangle_\phi$ in $\mathcal{L}_\phi$. Thus $(\mathcal{L}_\phi/N_\phi; \langle \cdot | \cdot \rangle_\phi)$ is a complex prehilbert space. Define a continuous linear representation $(\gamma, \mathcal{L}_\phi/N_\phi)$ of G by the prescription

$$\gamma(x) : p(f) \rightsquigarrow p(\varepsilon_x \star f).$$

It is easy to verify that

$$\langle \gamma(x) p(f) | \gamma(x) p(g) \rangle_\phi = \langle p(f) | p(g) \rangle_\phi$$

holds for all pairs $(f,g) \in \mathcal{L}_\phi \times \mathcal{L}_\phi$ and that we have

$$\langle \gamma(x) p(\phi) | p(\phi) \rangle_\phi = \phi(x)$$

for all $x \in G$. Let $\mathcal{H}_0$ denote the completion of the complex prehilbert space $(\mathcal{L}_\phi/N_\phi; \langle \cdot | \cdot \rangle_\phi)$ and extend the continuous linear representation $(\gamma, \mathcal{L}_\phi/N_\phi)$ of G to a continuous, unitary, linear representation $(U_0, \mathcal{H}_0)$ of G. If $f_0 \in \mathcal{H}_0$ denotes the vector that corresponds to $p(\phi) \in \mathcal{L}_\phi/N_\phi$ we have $\phi = c_{U_0, f_0, f_0}$, as stated. —

Remarks.

1. Lemma 1.13 shows that the function $\phi \in \mathfrak{P}(G)$ determines the continuous, unitary, linear representation $(U,\mathcal{H})$ of Theorem 1.21 up to unitary isomorphy. Indeed, if $(U_1,\mathcal{H}_1)$ denotes a second continuous, unitary, linear representation of G and $f_1 \in \mathcal{H}_1$ a cyclic vector with respect to $(U_1,\mathcal{H}_1)$ such that

$$\phi = c_{U_1,f_1,f_1}$$

holds then the $\mathbb{C}$-linear mapping

$$S_0 : \mathcal{L}_\phi/N_\phi \ni p(\mu*\phi) \rightsquigarrow U_1^1(\mu)f_1 \in \mathcal{H}_1$$

is isometric. Indeed we have

$$\begin{aligned}\|U_1^1(\mu)f_1\|^2 &= \langle U_1^1(\mu^* * \mu)f_1 | f_1\rangle \\ &= \langle \phi,\mu^* * \mu\rangle \\ &= \int_G \bar{\mu} * \phi(x)d\mu(x) \\ &= \|p(\mu*\phi)\|^2 .\end{aligned}$$

Thus S_0 extends to a unitary isomorphism $S \in \mathfrak{R}_G(U_0,U_1)$ of the complex Hilbert space $\mathcal{H}_0$ onto the complex Hilbert space $\mathcal{H}_1$ and satisfies $S(f_0) = f_1$.

2. Every function $\phi \in \mathfrak{P}(G)$ defines as an element of $L^\infty(G)$ a continuous positive linear form $\psi \rightsquigarrow \langle\psi,\phi\rangle = \int_G \psi(x)\phi(x)dx$ on the involutory complex Banach algebra $L^1(G)$. Indeed, by Theorem 1.20 we have

$$\langle\psi^* * \psi , \phi\rangle \geqq 0$$

for every $\psi \in L^1(G)$ and by Theorem 1.21 we get for the integrated form of $(U_0,\mathcal{H}_0)$ the identity

$$\langle\psi,\phi\rangle = c_{U_0^1,f_0,f_0}(\psi)$$

where $f_0 \in \mathcal{H}_0$ is a cyclic vector with respect to $(U_0, \mathcal{H}_0)$.

3. Let $\mathfrak{M}(G)$ denote the complex vector space of measures on the unimodular, locally compact, topological group G. Then $\mathfrak{M}(G)$ is the topological dual of the complex vector space $\mathcal{K}(G)$ under its canonical inductive limit topology. Observe that $\mathcal{K}(G)$ is an everywhere dense involutory subalgebra of $L^1(G)$. A complex measure $\mu \in \mathfrak{M}(G)$ is said to be *of positive type on* G if

$$\langle \psi^* * \psi, \mu \rangle \geqq 0$$

holds for every function $\psi \in \mathcal{K}(G)$. In the case $\mu = \phi . dx$ where $\phi \in \mathfrak{C}(G)$ this definition implies that μ is a measure of positive type on G if and only if $\phi \in \mathfrak{P}(G)$. Let G denote a unimodular real Lie group and $\mathfrak{D}(G)$ the vector subspace of $\mathcal{K}(G)$ of infinitely differentiable complex-valued functions on G with compact support. Let $\mathfrak{D}(G)$ be endowed with its canonical inductive limit topology. A complex distribution $T \in \mathfrak{D}'(G)$ is said to be *of positive type on* G if

$$\langle \psi^* * \psi, T \rangle \geqq 0$$

holds for every local test function $\psi \in \mathfrak{D}(G)$.

Let $(U, \mathcal{H})$ denote a continuous, unitary, linear representation of G in the complex Hilbert space $\mathcal{H}$. For $k \in \mathbb{N} \cup \{+\infty\}$ let $\mathcal{H}^{-k}$ be the *strong topological antidual* of the locally convex topological vector space $\mathcal{H}^k$ of k-times differentiable vectors for $(U, \mathcal{H})$; cf. 1.2, Remark 3. Then we get the ascending filtration of continuous linear injections

$$\mathcal{H}^0 \hookrightarrow \ldots \hookrightarrow \mathcal{H}^{-k} \hookrightarrow \mathcal{H}^{-(k+1)} \hookrightarrow \ldots \hookrightarrow \mathcal{H}^{-\infty}$$

where $\mathcal{H}^0 = \mathcal{H}$ and the natural continuous *sesquilinear* form $\langle \cdot | \cdot \rangle$ on $\mathcal{H}^\infty \times \mathcal{H}^{-\infty}$ extends the restriction of the scalar product of $\mathcal{H}$ onto $\mathcal{H}^\infty \times \mathcal{H}^\infty$. For every $\phi \in \mathfrak{D}(G)$ the $\mathbb{C}$-linear mapping

$$U^1(\phi) : \mathcal{H}^{-\infty} \to \mathcal{H}^\infty$$

is continuous with respect to the inductive limit topology of $\mathcal{H}^{-\infty} = \bigcup_{k \in \mathbb{N}} \mathcal{H}^{-k}$ and the projective limit topology of $\mathcal{H}^\infty = \bigcap_{k \in \mathbb{N}} \mathcal{H}^k$. For every pair $(f, g) \in \mathcal{H}^{-\infty} \times \mathcal{H}^{-\infty}$ the continuous linear form

$$c_{U^1,f,g} : \mathcal{D}(G) \ni \phi \rightsquigarrow \langle U^1(\phi)f|g\rangle \in \mathbb{C}$$

defines the *coefficient distribution of* $(U^1,\mathcal{H})$ *with respect to* $(f,g) \in \mathcal{H}^{-\infty}\times\mathcal{H}^{-\infty}$.

A complex distribution $T \in \mathcal{D}'(G)$ is *of positive type on* G if and only if there exists a continuous, unitary, linear representation $(U_0,\mathcal{H}_0)$ of G and a distribution $f_0 \in \mathcal{H}_0^{-\infty}$ such that the identity

$$T = c_{U_0^1,f_0,f_0}$$

holds. The set $\{U_0^1(\phi)f_0|\phi\in \mathcal{D}(G)\}$ is everywhere dense in $\mathcal{H}$ and T determines $(U_0,\mathcal{H}_0)$ up to unitary isomorphy.

1.22 Let G denote a locally compact topological group. Fix a left Haar measure dx on G and form the complex Hilbert space $L^2(G) = L^2(G;dx)$. The linear left action $G \times L^2(G) \ni (x,f) \to \gamma_G(x)f \in L^2(G)$ of G on $L^2(G)$ given by the convolution product

$$\gamma_G(x) : L^2(G) \ni f \rightsquigarrow \varepsilon_x * f \in L^2(G)$$

for all $x \in G$ (cf. 1.17) defines a continuous, unitary, linear representation $(\gamma_G,L^2(G))$ of G. It is called the *left regular representation of* G *in* $L^2(G)$. Obviously we have

$$\gamma_G(x)f : G \ni y \rightsquigarrow f(x^{-1}y) \in \mathbb{C}$$

for any element f of the stable everywhere dense vector subspace $\mathcal{K}(G)$ of $L^2(G)$. Thus the linear left G-action $G \times \mathcal{K}(G) \ni (x,f) \rightsquigarrow \gamma_G(x)f \in \mathcal{K}(G)$ on $\mathcal{K}(G)$ arises by lifting the homeomorphisms $\gamma_G(x) : G \ni y \rightsquigarrow xy \in G$ from G to $\mathcal{K}(G)$ so that the identity $\gamma_G(x)f(\gamma_G(x)y) = f(y)$ holds for all elements x and y in G. Obviously

$$\gamma_G^1(\mu)f = \mu * f$$

holds for all $\mu \in \mathfrak{M}^1(G)$ and $f \in L^2(G)$ and in particular we have

$$\gamma_G^1(g)f = g * f$$

for all elements $g \in L^1(G)$. If G is a *unimodular*, locally compact, topological group, the situation is similar with the *right regular representation*

$(\delta_G, L^2(G))$ *of* G *in* $L^2(G)$ given by the convolution product

$$\delta_G(y) : L^2(G) \ni f \rightsquigarrow f * \varepsilon_{y^{-1}} \in L^2(G)$$

for $y \in G$. Indeed, $(\delta_G, L^2(G))$ is a continuous, unitary, linear representation of G and $f \rightsquigarrow \check{f}$ is a unitary isomorphism of $(\gamma_G, L^2(G))$ onto $(\delta_G, L^2(G))$. Both regular representations may be combined in order to obtain the continuous, unitary, linear representation $(\gamma_G \times \delta_G, L^2(G))$ of $G \times G$ via the prescription

$$\gamma_G \times \delta_G : (x,y) \rightsquigarrow \gamma_G(x) \circ \delta_G(y).$$

The third main problem of harmonic analysis can be described as follows: Given a locally compact topological group G, determine the isomorphy classes in the unitary dual $\hat{G}$ that have representatives contained in the left regular representation $(\gamma_G, L^2(G))$ of G.

1.23 Let G be a topological group and $\phi \in \mathfrak{P}(G)$. The function ϕ is said to be a *minimal* continuous function of positive type on G if any additive decomposition

$$\phi = \phi_1 + \phi_2$$

of ϕ with functions $\phi_\alpha \in \mathfrak{P}(G)$, $\alpha \in \{1,2\}$, necessarily implies

$$\phi_\alpha = \zeta_\alpha \cdot \phi$$

with scalars $\zeta_\alpha \in \mathbb{C}$ and $\zeta_1 + \zeta_2 = 1$.

1.24 Theorem. Let G denote a topological group. The function $\phi \in \mathfrak{P}(G)$ is minimal on G if and only if there exists a topologically irreducible, continuous, unitary, linear representation $(U_0, \mathcal{H}_0)$ of G and a vector $f_0 \in \mathcal{H}_0$ such that

$$\phi = c_{U_0, f_0, f_0}$$

holds.

Proof. In view of Theorem 1.21 there exists a continuous, unitary, linear representation $(U_0, \mathcal{H}_0)$ of G and a cyclic vector $f_0 \in \mathcal{H}_0$ with respect to

$(U_o, \mathcal{H}_o)$ such that $\phi = c_{U_o, f_o, f_o}$ holds. Suppose that $\phi \in \mathfrak{P}(G)$ is minimal on G and that there is a closed vector subspace $\mathcal{L}$ of $\mathcal{H}_o$ which is stable under U_o. Let $P_{\mathcal{L}}$ denote the orthoprojector of $\mathcal{H}_o$ onto $\mathcal{L}$ and let $P_{\mathcal{L}} f_o = f_1 \in \mathcal{L}$, $P_{\mathcal{L}^\perp} f_o = f_2 = f_o - f_1 \in \mathcal{L}^\perp$. Define the functions

$$\phi_1 = c_{U_o, f_1, f_1} = c_{U_o, f_1, f_o} \in \mathfrak{P}(G),$$

and

$$\phi_2 = c_{U_o, f_2, f_2} = c_{U_o, f_2, f_o} \in \mathfrak{P}(G).$$

It follows $\phi = \phi_1 + \phi_2$ and therefore $\phi_1 - \zeta_1 \phi = 0$ where $\zeta_1 \in \mathbb{C}$ is a suitable scalar. Consequently we have

$$c_{U_o, f_o, f_1 - \zeta_1 f_o} = 0.$$

Since $f_o \in \mathcal{H}_o$ is a cyclic vector with respect to $(U_o, \mathcal{H}_o)$ it follows $f_1 = \zeta_1 f_o$. In the case $\zeta_1 = 0$ we have $\mathcal{L} = \{0\}$, and in the case $\zeta_1 \neq 0$ of course $\mathcal{L} = \mathcal{H}_o$. Thus $(U_o, \mathcal{H}_o)$ is a topologically irreducible representation of G.

Conversely, suppose that $(U_o, \mathcal{H}_o)$ is a topologically irreducible, continuous, unitary, linear representation of G which admits a vector $f_o \in \mathcal{H}_o$ such that $\phi = c_{U_o, f_o, f_o}$. In view of 1.21, Remark 1 we may assume that $\mathcal{H}_o$ is the completion of the complex prehilbert space $(\mathcal{L}_\phi / N_\phi ; \langle \cdot | \cdot \rangle_\phi)$ and U_o is the extension of γ to $\mathcal{H}_o$. Assume the decomposition $\phi = \phi_1 + \phi_2$ with $\phi_\alpha \in \mathfrak{P}(G)$, $\alpha \in \{1,2\}$. The positive Hermitean sesquilinear form $\langle \cdot | \cdot \rangle_{\phi_1}$ is continuous on the prehilbert space $(\mathcal{L}_\phi / N_\phi ; \langle \cdot | \cdot \rangle_\phi)$ since we have the estimate

$$|\langle p(f) | p(g) \rangle_{\phi_1}| \leq \|p(f)\|_\phi \cdot \|p(g)\|_\phi$$

for all pairs $(f,g) \in \mathcal{L}_\phi$. Consequently $\langle \cdot | \cdot \rangle_{\phi_1}$ extends to $\mathcal{H}_o \times \mathcal{H}_o$ and there exists a continuous Hermitean positive linear operator $K : \mathcal{H}_o \to \mathcal{H}_o$ such that

$$\langle f_o | K g_o \rangle_\phi = \langle f_o | g_o \rangle_{\phi_1}$$

holds for all pairs $(f_o, g_o) \in \mathcal{H}_o \times \mathcal{H}_o$. If the complex Hilbert space $(\mathcal{H}_o, \langle \cdot | \cdot \rangle_{\phi_1})$ is identified with its topological antidual then K forms the

Schwartz kernel of the Hilbert subspace $(\mathcal{H}_0, \langle\cdot|\cdot\rangle_\phi)$ of $(\mathcal{H}_0, \langle\cdot|\cdot\rangle_{\phi_1})$. Obviously the operators $U_0(x)$ preserve $\langle\cdot|\cdot\rangle_{\phi_1}$ for all $x \in G$. It follows $K \in \mathfrak{R}_G(U_0)$ and therefore we have $K = \zeta_1 \mathrm{id}_{\mathcal{H}_0}$ $(\zeta_1 \in \mathbb{C})$ by the Corollary of Theorem 1.6. Consequently $\phi_1 = \zeta_1\phi$, i.e., $\phi \in \mathfrak{P}(G)$ is minimal on G. —

References

Gross, K.I. : On the evolution of noncommutative harmonic analysis. Amer. Math. Monthly 85 (1978), 525-548.

Mackey, G.W. : Origins and early history of the theory of unitary group representations. In: Representation theory of Lie groups, pp. 5-19. London Mathematical Society Lecture Note Series, Vol. 34. Cambridge University Press, Cambridge 1979.

Mackey, G.W. : Harmonic analysis as the exploitation of symmetry - a historical survey. Bull. (New Series) Amer. Math. Soc. 3 (1980), 543-698.

Stewart, J. : Positive definite functions and generalizations, an historical survey. Rocky Mountain J. Math. 6 (1976), 409-434. Also in: Reproducing Hilbert spaces. H.L. Weinert, ed., pp. 74-99. Hutchinson Ross Publishing Company, Stroudsburg, PA, 1982.

2 The unitary inducing procedure

2.1 Let G denote a locally compact topological group and H a *closed* subgroup of G. Then H acts continuously on the *right* on G. Indeed, for any element $h \in H$ the right translation by $h^{-1} \in H$ in G defined according to

$$\delta_G(h) : G \ni x \rightsquigarrow xh^{-1} \in G$$

is a homeomorphism of G into itself, we have $\delta_G(1_G) = id_G$, $\delta_G(hk) = \delta_G(h) \circ \delta_G(k)$ for all pairs $(h,k) \in H \times H$, and the mapping defined via the prescription

$$H \times G \ni (h,x) \rightsquigarrow \delta_G(h)x \in G$$

is continuous with respect to the product topology of the direct product group $H \times G$ and the given group topology of G. The orbit space associated with this *right* action of H on G is the set G/H of *left* cosets $\dot{x} = \delta_G(H)x = x.H$ of H in G. If the orbit space G/H is endowed with its natural Hausdorff topology, i.e., with the quotient by H of the topology of G, then the locally compact topological space G/H is said to be the *homogeneous space of left cosets of* H *in* G. The homogeneous space $H\backslash G$ of *right* cosets $\dot{x} = \gamma_G(H)x = H.x$ of H in G is analogously defined.

Let $p : G \to G/H$ denote the canonical surjection. Then the map

$$p : G \ni x \rightsquigarrow \dot{x} = \delta_G(H)x = x.H \in G/H$$

is *open* and *continuous* with respect to the group topology of G and the quotient topology of G/H. The troup G acts continuously on the *left* on G/H. Indeed, if $\gamma_G(x) : G \ni y \rightsquigarrow xy \in G$ denotes the left translation by $x \in G$ in G (cf. 1.22) there exists a unique continuous mapping $\gamma_{G/H}(x) : G/H \to G/H$ which makes the following diagram commutative for all $x \in G$.

$$\begin{array}{ccc} G & \xrightarrow{\gamma_G(x)} & G \\ p\downarrow & & \downarrow p \\ G/H & \xrightarrow[\gamma_{G/H}(x)]{} & G/H \end{array}$$

Thus we have

$$\gamma_{G/H}(x)(y.H) = (xy).H$$

for all elements $y \in G$ and the continuous *left* action

$$(x,\dot{y}) \rightsquigarrow \gamma_{G/H}(x)\dot{y}$$

of G on the homogeneous space G/H is transitive.

The natural *right* action

$$(x,\dot{y}) \rightsquigarrow \delta_{H\backslash G}(x)\dot{y}$$

on G on the *homogeneous space* $H\backslash G$ *of right cosets of* H *in* G has similar properties provided $\dot{y} = \gamma_G(H)y = H.y \in H\backslash G$ for $y \in G$.

Choose a *left* Haar measure dh on H and define for every function $\phi \in \mathbb{K}(G)$ and for any element $x \in G$ the *orbit mean*

$$\phi^b(x) = \int_H \delta_G(h)\phi(x)dh = \int_H \phi(xh)dh$$

which exists since the function $h \rightsquigarrow \delta_G(h)\phi(x) = \phi(xh)$ belongs to the vector space $\mathbb{K}(H)$ for all $x \in G$. The complex valued function ϕ^b which, of course, depends upon the choice of dh, is continuous on G and invariant under the *right* translations by elements $h \in H$ in G, i.e., $\delta_G(h)\phi^b = \phi^b$, since dh is a left Haar measure on H. It follows that there exists a unique continuous function $\Phi: G/H \to \mathbb{C}$ so that the factorization $\phi^b = \Phi \circ p$ holds. Thus the following diagram commutes.

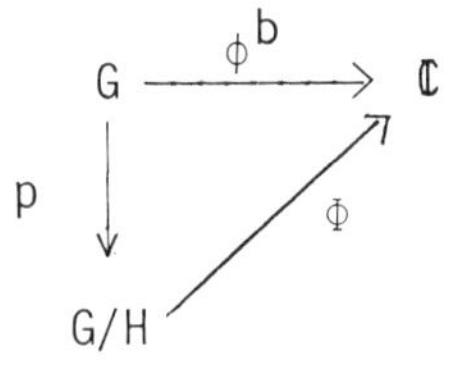

Since $\text{Supp}(\Phi) \subseteq p(\text{Supp}(\phi))$ is compact in G/H we have $\Phi \in \mathbb{K}(G/H)$. By identifying ϕ^b with Φ the mapping

$$\mathbb{K}(G) \ni \phi \rightsquigarrow \phi^b \in \mathbb{K}(G/H)$$

is well defined and $\mathbb{C}$-linear. Let $\mathbb{K}_+(G)$ and $\mathbb{K}_+(G/H)$ denote the convex cones of functions ≥ 0 in the complex vector spaces $\mathbb{K}(G)$ and $\mathbb{K}(G/H)$, respectively.

Then we have the following technical result which will be useful in the sequel.

2.2 Lemma. The mapping

$$\mathbb{K}_+(G) \ni \phi \rightsquigarrow \phi^b \in \mathbb{K}_+(G/H)$$

is surjective.

Proof Let $\Psi \in \mathbb{K}_+(G/H)$ and $L = \mathrm{Supp}(\Psi)$. There exists a compact subset K of G so that $p(K) = L$ since a straightforward covering argument shows that every compact subset of G/H is the image under the canonical surjection p of a compact subset of G. Choose a function $\phi \in \mathbb{K}_+(G)$ satisfying

$$\phi(x) > 0 \text{ for } x \in K.$$

Then $\phi^b \in \mathbb{K}_+(G/H)$ and $\phi^b(\dot{x}) > 0$ for $\dot{x} \in L$. Define the function

$$\Phi(\dot{x}) = \begin{cases} \dfrac{\Psi(\dot{x})}{\phi^b(\dot{x})} & \text{whenever } \dot{x} \in L, \\[2ex] 0 & \text{whenever } \dot{x} \notin L. \end{cases}$$

Then $\Phi \in \mathbb{K}_+(G/H)$ and $\Psi = \Phi \cdot \phi^b$ on G/H. The product $\phi \cdot (\Phi \circ p)$ belongs to $\mathbb{K}_+(G)$. By virtue of the fact that $\Phi \circ p$ is invariant under the right translations $\delta_G(h)$ by elements $h \in H$ in G we have

$$(\phi.(\Phi \circ p))^b = \phi^b.\Phi = \Psi.$$

This identity proves the lemma. —

Corollary. The $\mathbb{C}$-linear mapping

$$\mathbb{K}(G) \ni \phi \rightsquigarrow \phi^b \in \mathbb{K}(G/H)$$

is a continuous surjection with respect to the canonical inductive limit topologies of the complex vector spaces $\mathbb{K}(G)$ and $\mathbb{K}(G/H)$, respectively.

2.3 Suppose that there exists a positive measure $\mu \neq 0$ on G/H which is invariant under the natural *left* action $(x,\dot{y}) \rightsquigarrow \gamma_{G/H}(x)\dot{y}$ of G on G/H, i.e.,

$\gamma_{G/H}(x)\mu = \mu$ for all $x \in G$. Then the continuous linear form

$$\mathbb{K}(G) \ni \phi \rightsquigarrow \int_{G/H} \phi^b(\dot{x})d\mu(\dot{x}) \in \mathbb{C}$$

on the complex, locally convex, topological vector space $\mathbb{K}(G)$ is a left Haar measure dx on G. In this case we have the identity

$$\int_G \phi(x)dx = \int_{G/H} (\int_H \phi(xh)dh)d\mu(\dot{x})$$

for all functions $\phi \in \mathbb{K}(G)$. It should be observed that the intergral $\int_H \phi(xh)dh$ on the right hand side of this identity has to be considered as a function of the left cosets $\dot{x} = x.H = p(x)$ of H in G and *not* as a function of the elements $x \in G$. In the present case the measure

$$d\mu = \frac{dx}{dh}$$

on G/H which we assumed to exist, is called to be the *quotient* of the left Haar measure dx of G by the left Haar measure dh of H. By definition of the right modular function Δ_G of G (cf. 1.16) we have the identity

$$\delta_G(y)dx = \Delta_G(y)dx$$

for all $y \in G$. Now suppose $y \in H$. Then we have by the remark above

$$\begin{aligned} \Delta_G(y) \int_G \phi(x)dx &= \int_{G/H} (\int_H \phi(xhy^{-1})dh)dx \\ &= \Delta_H(y) \int_{G/H} (\int_H \phi(xh)dh)dx \\ &= \Delta_H(y) \int_G \phi(x)dx \end{aligned}$$

for all functions $\phi \in \mathbb{K}(G)$. It follows that the identity

$$\Delta_G|H = \Delta_H$$

is a necessary condition for the existence of a positive measure $\mu \neq 0$ on G/H such that $\gamma_{G/H}(x)\mu = \mu$ holds for all elements $x \in G$. The next theorem will show that this condition is also sufficient.

2.4 Theorem. Let G denote a locally compact topological group and H a closed subgroup of G. There exists a positive measure $\mu \neq 0$ on the homogeneous space G/H that is invariant under the natural left action $(x,y) \rightsquigarrow \gamma_{G/H}(x)\dot{y}$ of G on G/H if and only if the right modular functions of G and H satisfy the identity

$$\Delta_G | H = \Delta_H.$$

In this case there exists a left Haar measure dx on G and a left Haar measure dh on H such that

$$d\mu = \frac{dx}{dh}$$

holds. If we choose arbitrarily one of the Haar measures dx and dh on G and H, respectively, then μ determines uniquely the other one.

Proof. Suppose that the identity $\Delta_G | H = \Delta_H$ holds and choose a *left* Haar measure dh on H. For any pair ϕ,ψ of functions in $\mathbb{K}(G)$ let ϕ^b and ψ^b be the orbit means of ϕ and ψ, respectively, with respect to dh. An application of the Lebesgue-Fubini theorem then shows the identities

$$\begin{aligned}\int_G \phi(x)\psi^b(x)dx &= \int_G \phi(x) \left(\int_H \psi(xh)dh\right)dx \\ &= \int_H \left(\int_G \phi(x)\psi(xh)dx\right)dh \\ &= \int_H \left(\int_G \phi(xh^{-1})\psi(x)\ \Delta_H(h)^{-1}dx\right)dh \\ &= \int_G \psi(x) \left(\int_H \phi(xh^{-1})\Delta_H(h)^{-1}dh\right)dx \\ &= \int_G \psi(x) \left(\int_H \phi(xh)dh\right)dx \\ &= \int_G \phi^b(x)\psi(x)dx.\end{aligned}$$

Let a function $\Phi \in \mathbb{K}(G/H)$ be given. By virtue of the Corollary of Lemma 2.2 there exists a function $\phi \in \mathbb{K}(G)$ such that $\phi^b = \Phi$. We will show that the linear form

$$\mu: \mathbb{K}(G/H) \ni \Phi \rightsquigarrow \int_G \phi(x)dx \in \mathbb{C}$$

is well defined. Indeed, let $\rho \in \mathbb{K}(G/H)$ denote another function such that $\rho^b = \Phi$. Then we have $(\phi-\rho)^b = 0$ and therefore by the preceding reasoning $\int_G (\phi-\rho)(x)\psi^b(x)dx = 0$ for all functions $\psi \in \mathbb{K}(G)$. Another application of the Corollary of Lemma 2.2 shows that there exists a function $\psi \in \mathbb{K}(G)$ satisfying $\psi^b|\mathrm{Supp}(\phi-\rho) = 1$. Hence $\int_G \rho(x)dx = \int_G \phi(x)dx$.

In the case $\Phi \in \mathbb{K}_+(G/H)$ we may choose $\phi \in \mathbb{K}_+(G)$ such that $\phi^b = \Phi$ by Lemma 2.2. Thus μ is a continuous linear form on the complex, locally convex, topological vector space $\mathbb{K}(G/H)$, hence a positive measure $\mu \in \mathfrak{M}(G/H)$ different from 0 that satisfies

$$\int_G \phi(x)dx = \int_{G/H} \left(\int_H \phi(xh)dh\right)d\mu(\dot{x})$$

for all functions $\phi \in \mathbb{K}(G)$. Consequently μ is invariant under the natural left action $(x,\dot{y}) \rightsquigarrow \gamma_{G/H}(x)\dot{y}$ of G on G/H. —

<u>Corollary</u>. Let G be a unimodular, locally compact, topological group and let $H \subseteq G$ be a closed subgroup so that H is unimodular - then the homogeneous space G/H admits a positive measure $\neq 0$ invariant under the natural G-action on G/H.

In the case when H is a closed *normal* subgroup of the locally compact topological group G, the quotient by H of the topology of G is compatible with the quotient group structure of G/H, and the quotient $d\mu = \frac{dx}{dh}$ of the left Haar measure dx of G by the left Haar measure dh of H is a left Haar measure μ of the locally compact topological group G/H. Two of the three Haar measures $\{dx,dh,d\mu\}$ on G, H, and G/H, respectively, determine the third one uniquely. The preceding theorem yields the identity $\Delta_G|H = \Delta_H$. In particular, if G is unimodular then H has the same property.

2.5 Let G denote a locally compact topological group and H a closed subgroup of G such that there exists a positive measure $\mu \neq 0$ on the homogeneous space G/H that is invariant under the natural left action $(x,\dot{y}) \rightsquigarrow \gamma_{G/H}(x)\dot{y}$ of G on G/H. For any $\Phi \in \mathbb{K}(G/H)$ and any element $x \in G$ define the lifted left action of G on Φ in the usual way according to the prescription

$$\gamma_{G/H}(x)\Phi : \dot{y} \rightsquigarrow \Phi(\gamma_{G/H}(x^{-1})\dot{y}).$$

Then $\gamma_{G/H}(x)$ defines for every element $x \in G$ an isometric endomorphism of the vector subspace $\mathbb{K}(G/H)$ of the complex Hilbert space $L^2(G/H) = L^2(G/H;\mu)$ and $\gamma_{G/H}$ extends to a continuous unitary linear representation of G in $L^2(G/H)$. The pair $(\gamma_{G/H}, L^2(G/H))$ is said to be the *left quasi-regular representation of* G *in* $L^2(G/H)$. It reduces to the *left regular representation* $(\gamma_G, L^2(G))$ of G when $H = \{1_G\}$ is the trivial subgroup of G (cf. Section 1.22).

2.6 Let G be a locally compact topological group and K,H two closed subgroups of G such that the following tower of continuous canonical monomorphisms

$$K \hookrightarrow H \overset{j}{\hookrightarrow} G$$

arises. Let dx be a left Haar measure on G and suppose that there exists a positive measure $\beta \neq 0$ on the homogeneous space G/K that is invariant under the natural left action $(x, y.K) \rightsquigarrow \gamma_{G/K}(x)y.K$ of G on G/K and a positive measure $\lambda \neq 0$ on the homogeneous space G/H that is invariant under the natural left action $(x,y) \rightsquigarrow \gamma_{G/H}(x)\dot{y}$ of G on G/H where $\dot{y} = \delta_G(H)y = y.H$ for $y \in G$. By virtue of Theorem 2.4 we have

$$\Delta_G | K = \Delta_H | K = \Delta_K$$

so that the homogeneous space H/K admits a positive measure $\alpha \neq 0$ that is invariant under the natural left action $(h, 1.K) \rightsquigarrow \gamma_{H/K}(h)(1.K)$ of H on H/K. Choose left Haar measures dk on K and dh on H such that the identities

$$d\beta = \frac{dx}{dk}, \qquad d\lambda = \frac{dx}{dh}, \qquad d\alpha = \frac{dh}{dk}$$

hold. Let $q : G \ni y \rightsquigarrow y.K \in G/K$, $p : G \ni y \rightsquigarrow \dot{y} = y.H \in G/H$, and $q_0 : H \ni 1 \rightsquigarrow 1.K \in H/K$ denote the continuous canonical surjections. By forming the quotient of the continuous canonical injection j of H in G there exists a unique continuous mapping $\omega: H/K \to G/K$ which makes the following diagram commutative:

$$\begin{array}{ccc} H & \overset{j}{\hookrightarrow} & G \\ {\scriptstyle q_0}\downarrow & & \downarrow{\scriptstyle q} \\ H/K & \underset{\omega}{\longrightarrow} & G/K \end{array}$$

The mapping ω is injective and closed. Hence ω defines a homeomorphism of

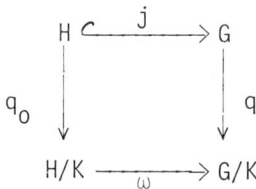

H/K onto the closed subset q(H) of G/K. Thus we may identify the homogeneous space H/K with q(H) and the mapping q_0 with the restriction $q|H$. In this case the measure α on H/K is identified with the image $\omega(\alpha) \in \mathfrak{M}(G/K)$ of α under ω that is concentrated on the closed subset q(H) of G/K.

For every element $x \in G$ consider the measure $\gamma_{G/K}(x) \circ \omega(\alpha)$ on the homogeneous space G/K. For all $h \in H$ we have

$$\gamma_{G/K}(xh) \circ \omega(\alpha) = \gamma_{G/K}(x) \circ \omega(\gamma_{H/K}(h)\alpha) = \gamma_{G/K}(x) \circ \omega(\alpha),$$

hence

$$\delta_{G/K}(h)(\gamma_{G/K}(x) \circ \omega(\alpha)) = \gamma_{G/K}(x) \circ \omega(\alpha).$$

Thus, by passing to the quotient, the mapping $G \ni x \rightsquigarrow \gamma_{G/K}(x) \circ \omega(\alpha) \in \mathfrak{M}(G/K)$ gives rise to a mapping

$$G/H \ni \dot{x} \rightsquigarrow (\frac{dh}{dk})_{\dot{x}} \in \mathfrak{M}(G/K).$$

It follows

2.7 Theorem. Retain the above notations and assumptions - then the transitivity of invariant measures

$$\frac{dx}{dk} = \int_{G/H} (\frac{dh}{dk})_{\dot{x}} \, d\lambda(\dot{x})$$

holds.

Proof. Let $\phi \in \mathbb{K}(G)$ be arbitrary and form the orbit mean $\phi^b \in \mathbb{K}(G/K)$ according to the prescription (cf. Section 2.1)

$$\phi^b(q(x)) = \int_K \phi(xk)dk$$

for all elements $x \in G$. Since the linear mapping $\mathbb{K}(G) \ni \phi \rightsquigarrow \phi^b \in \mathbb{K}(G/K)$ is surjective by the Corollary of Lemma 2.2, it will be sufficient to test the identity of transitivity by means of the functions $\{\phi^b | \phi \in \mathbb{K}(G)\}$. The left hand side of the asserted identity gives (cf. Section 2.3)

$$\int_{G/K} \phi^b \frac{dx}{dk} = \langle \phi^b, \beta \rangle = \int_G \phi(x)dx.$$

The right hand side yields

$$\langle \phi^b, \int_{G/H} (\tfrac{dh}{dk})_{\dot{x}}\, d\lambda(\dot{x})\rangle = \int_{G/H} \langle \phi^b, (\tfrac{dh}{dk})_{\dot{x}}\rangle\, d\lambda(\dot{x})$$

where the integrand of the right hand integral admits the following form

$$\begin{aligned}
\langle \phi^b, (\tfrac{dh}{dk})_{\dot{x}}\rangle &= \langle \phi^b, \gamma_{G/K}(x) \circ \omega(\alpha)\rangle \\
&= \int_{H/K} \phi^b(\gamma_{G/K}(x)\dot{h})d\alpha(\dot{h}) \\
&= \int_{H/K} \phi^b(q(xh))d\alpha(\dot{h}) \\
&= \int_{H/K} (\int_K \phi(xhk)dk)d\alpha(\dot{h}) \\
&= \int_H \phi(xh)dh.
\end{aligned}$$

Hence

$$\langle \phi^b, \int_{G/H} (\tfrac{dh}{dk})_{\dot{x}}\, d\lambda(\dot{x})\rangle = \int_{G/H} (\int_H \phi(xh)dh)d\lambda(\dot{x}) = \int_G \phi(x)dx.$$

Since in the preceding reasoning the function $\phi \in \mathbb{K}(G)$ is chosen arbitrarily, the proof is complete. —

2.8 Let G denote a locally compact topological group and H a closed subgroup of G. As our first hypothesis we shall suppose that G and H are unimodular. If $p : G \ni x \leadsto \dot{x} \in G/H$ denotes the continuous canonical surjection, a *section* s *of* G *fibred by* H is a mapping $G/H \ni \dot{x} \leadsto s(\dot{x}) \in G$ such that

$$p \circ s = id_{G/H}.$$

As our second hypothesis on G and H we shall suppose that there exists a *continuous* section s of G fibred by H. Then $(h,\dot{x}) \leadsto s(\dot{x})h$ is a *homeomorphism* of $H \times (G/H)$ onto G and its inverse is given by the assignment $x \leadsto (s(\dot{x})^{-1}x,\dot{x})$. For the unitary inducing procedure which we are going to describe in the present section, both of the hypotheses are actually superfluous. However, they are convenient to avoid some subtleties of the unitary inducing construction and they are automatically satisfied in the case of simply connected, nilpotent, real Lie groups G.

Suppose that $(U_0, \mathcal{H}_0)$ denotes a continuous, unitary, linear representation of

H in the complex Hilbert space ($\mathcal{H}_0$; $\langle\cdot|\cdot\rangle_0$). Then we denote by $\mathcal{K}(G/H,\mathcal{H}_0)$ the complex vector space of all continuous mappings $\phi:G \to \mathcal{H}_0$ that satisfy with respect to the natural *right* action $(h,x) \rightsquigarrow \delta_G(h)x$ of H on G the *covariance condition*

$$\delta_G(h)\phi = U_0(h^{-1})\phi$$

for all elements $h \in H$, and whose support $\mathrm{Supp}(\phi)$ is the saturation with respect to the natural right H-action on G of a *compact* subset of G. For any pair ϕ,ψ of mappings in $\mathcal{K}(G/H,\mathcal{H}_0)$, the continuous function

$$G \ni x \rightsquigarrow \langle\phi(x)|\psi(x)\rangle_0 \in \mathbb{C}$$

is invariant under the natural right H-action on G and its support is the saturation of a compact subset of G. Since $\Delta_G|H = \Delta_H = 1$, the quotient $d\mu = \frac{dx}{dh}$ of a left (hence right) Haar measure dx on G by a left (hence right) Haar measure dh on H defines a positive measure $\mu \neq 0$ on the homogeneous space G/H that is invariant under the natural left action $(x,\dot{y}) \rightsquigarrow \gamma_{G/H}(x)\dot{y}$ of G on G/H (Corollary of Theorem 2.4). Define on the complex vector space $\mathcal{K}(G/H,\mathcal{H}_0)$ the sesquilinear form

$$(\phi,\psi) \rightsquigarrow \langle\phi|\psi\rangle_\mu = \int_{G/H} \langle\phi(x)|\psi(x)\rangle_0 d\mu(\dot{x}).$$

Then $(\mathcal{K}(G/H,\mathcal{H}_0); \langle\cdot|\cdot\rangle_\mu)$ is a separated complex prehilbert space (with norm $\|\cdot\|_\mu$).

For every element $x \in G$ define the $\mathbb{C}$-linear mapping

$$\gamma_{G/H}(x) : \mathcal{K}(G/H,\mathcal{H}_0) \to \mathcal{K}(G/H,\mathcal{H}_0)$$

which assigns to every element $\phi \in \mathcal{K}(G/H,\mathcal{H}_0)$ the mapping

$$\gamma_{G/H}(x) : y \rightsquigarrow \phi(\gamma_{G/H}(x^{-1})\dot{y})$$

where $\dot{y} = \delta_G(H)y = y.H \in G/H$ for $y \in G$. Then $\gamma_{G/H}(x)$ is an isometric isomorphism of $(\mathcal{K}(G/H,\mathcal{H}_0); \langle\cdot|\cdot\rangle_\mu)$ onto itself which extends to a unique linear automorphism U(x) of the *completion* $\mathcal{H} = L^2(G/H,\mathcal{H}_0;\mu)$ of the complex prehilbert space $(\mathcal{K}(G/H,\mathcal{H}_0); \langle\cdot|\cdot\rangle_\mu)$

2.9 Theorem. Let H denote a unimodular closed subgroup of the unimodular, locally compact, topological group G such that there exists a continuous section of G fibred by H. Starting with the continuous, unitary, linear representation $(U_0,\mathcal{H}_0)$ of H, the preceding construction gives rise to a continuous, unitary, linear representation $(U,\mathcal{H})$ of G.

Proof. Let $x \in G$ and $\phi \in \mathcal{K}(G/H,\mathcal{H}_0)$ be arbitrary. Observe that $\mathcal{K}(G/H,\mathcal{H}_0)$ is an everywhere dense vector subspace of the complex Hilbert space $\mathcal{H} = L^2(G/H,\mathcal{H}_0;\mu)$. The function $G/H \ni \dot{y} \rightsquigarrow \|\gamma_{G/H}(x)\phi(y) - \phi(y)\|_0^2 \in \mathbb{R}_+$ belongs to the vector space $\mathcal{K}(G/H)$. Since we have

$$\|U(x)\phi-\phi\|_\mu^2 = \int_{G/H} \|\gamma_{G/H}(x)\phi(y) - \phi(y)\|_0^2 \, d\mu(\dot{y})$$

there exists for every $\varepsilon > 0$ a neighourhood V of 1_G in G such that

$$\|U(x)\phi-\phi\|_\mu < \varepsilon$$

for all $x \in V$. Thus the mapping $G \ni x \rightsquigarrow U(x)\phi \in \mathcal{K}(G/H,\mathcal{H}_0)$ is continuous at the neutral element 1_G of G, hence globally on G. A density argument then shows that for every element $\phi \in \mathcal{H}$ the mapping

$$G \ni x \rightsquigarrow U(x)\phi \in \mathcal{H}$$

is continuous. Thus $(U,\mathcal{H})$ is a continuous, unitary, linear representation of G in the complex Hilbert space $\mathcal{H}$. ——

The isomorphy class of the continuous, unitary, linear representation $(U,\mathcal{H})$ of G does not depend upon the specific choice of the measure μ on G/H. It also does not depend upon the specific choice of the continuous, unitary, linear representation $(U_0,\mathcal{H}_0)$ of H within its isomorphy class. We shall use the notation

$$(U,\mathcal{H}) = \mathrm{Ind}_H^G(U_0,\mathcal{H}_0)$$

and call $(U,\mathcal{H})$ the linear representation of G in the complex Hilbert space $\mathcal{H} = L^2(G/H,\mathcal{H}_0;\mu)$ *unitarily induced* by the given continuous, unitary, linear representation $(U_0,\mathcal{H}_0)$ of H in $\mathcal{H}_0$. The realization $(U,\mathcal{H})$ of $\mathrm{Ind}_H^G(U_0,\mathcal{H}_0)$ acts on the elements $\phi \in \mathcal{K}(G/H,\mathcal{H}_0)$ by means of the lifted *left* translations $\gamma_{G/H}(x)$ by elements $x \in G$.

Remarks.

1. It is sometimes convenient to formulate the unitary inducing procedure by

means of the homogeneous space $H\backslash G$ of *right* cosets of H in G. In this case the complex vector space $\mathcal{K}(H\backslash G,\mathcal{H}_0)$ is formed by the continuous mappings $\phi: G \to \mathcal{H}_0$ satisfying with respect to the natural *left* action $(h,x) \rightsquigarrow \gamma_G(h)x$ of H on G the *covariance condition*

$$\gamma_G(h)\phi = U_0(h^{-1})\phi$$

for all elements $h \in H$, and whose support $\mathrm{Supp}(\phi)$ is the saturation with respect to the natural *left* H-action on G of a compact subset of G. In the present situation the realization of $\mathrm{Ind}_H^G(U_0,\mathcal{H}_0)$ acts on the elements $\phi \in \mathcal{K}(H\backslash G,\mathcal{H}_0)$ by means of the lifted *right* translations $\delta_{H\backslash G}(x)$ by elements $x \in G$, i.e ,

$$\delta_{H\backslash G}(x)\phi \;:\; \dot{y} \rightsquigarrow \phi(\delta_{H\backslash G}(x^{-1})\dot{y})$$

where $\dot{y} = \gamma_G(H)y = H.y \in H\backslash G$ for $y \in G$.

2. The unitarily induced linear representation $(U,\mathcal{H}) = \mathrm{Ind}_H^G(U_0,\mathcal{H}_0)$ of G is topologically irreducible only if $(U_0,\mathcal{H}_0)$ is a topologically irreducible continuous, unitary, linear representation of H. Indeed, if $(U_\alpha,\mathcal{H}_\alpha)$ are continuous, unitary, linear representations of G for $\alpha \in \{1,2\}$, then $\mathrm{Ind}_H^G(U_1 \oplus U_2, \mathcal{H}_1 \oplus \mathcal{H}_2)$ and $\mathrm{Ind}_H^G(U_1,\mathcal{H}_1) \oplus \mathrm{Ind}_H^G(U_2,\mathcal{H}_2)$ are isomorphic, continuous, unitary, linear representations of G (cf. 1.4). Let χ_{U_0} denote the central character of $(U_0,\mathcal{H}_0)$ (see 1.6). Then we have

$$\mathrm{Ind}_H^G(U_0,\mathcal{H}_0)(z) = \chi_{U_0}(z)\mathrm{id}_{\mathcal{H}}$$

for all elements z of the centre C of G.

3. If $(U_0,\mathcal{H}_0)$ is a one-dimensional, continuous, unitary, linear representation of H, i.e., if $\mathcal{H}_0 = \mathbb{C}$ and U_0 is a continuous unitary character χ of H then $(U,\mathcal{H}) = \mathrm{Ind}_H^G(U_0,\mathcal{H}_0)$ is called to be a *monomial* representation of G (cf. 2.17 infra). In this case we denote the complex vector space $\mathcal{K}(G/H,\mathcal{H}_0)$ by $\mathcal{K}(G/H,\chi)$ and the complex Hilbert space $L^2(G/H,\mathcal{H}_0;\mu)$ by $L^2(G/H,\chi;\mu)$. The spaces $\mathcal{K}(H\backslash G,\chi)$ and $L^2(H\backslash G,\chi;\mu)$ are similarly defined. The covariance conditions of the functions $\phi \in \mathcal{K}(G/H,\chi)$ and $\psi \in \mathcal{K}(H\backslash G,\chi)$ take the form

$$\phi(xh) = \bar{\chi}(h)\phi(x), \quad \psi(hx) = \chi(h)\psi(x),$$

respectively, for all pairs $(h,x) \in H \times G$. In this case we will say that the function $\phi \in \mathbb{K}(G/H,\chi)$ *fibres over* χ *on the left* and $\psi \in \mathbb{K}(H\backslash G,\chi)$ *fibres over* χ *on the right.*

For $\phi \in \mathbb{K}(G)$ and $\psi \in \mathbb{K}(G/H,\chi)$ we obtain by a change of variables for the continuous linear mapping

$$U^1(\phi) : L^2(G/H,\chi;\mu) \to L^2(G/H,\chi;\mu)$$

the explicit form

$$(U^1(\phi)\psi) : \dot{x} \rightsquigarrow \int_{G/H} K_\phi(s(\dot{x}), s(\dot{y}))\,\psi(\dot{y})\,d\mu(\dot{y})$$

where the *kernel* $(G/H) \times (G/H) \ni (\dot{x},\dot{y}) \rightsquigarrow K_\phi(s(\dot{x}), s(\dot{y})) \in \mathbb{C}$ is continuous by the continuity of the section s of G fibred by H and the identity

$$K_\phi(x,y) = \int_H ((\check{\gamma}_G(x) \times \check{\delta}_G(y))\phi(h)).\chi(h)dh$$

for $(x,y) \in G \times G$.

Example. The left quasi-regular representation $(\gamma_{G/H}, L^2(G/H;\mu))$ is induced by the trivial representation $H \ni h \rightsquigarrow U_0(h) = \mathrm{id}_{\mathbb{C}}$ of H in the one-dimensional complex vector space $\mathbb{H}_0 = \mathbb{C}$; cf. Section 2.5 supra. In this case we have $\mathbb{K}(G/H,\mathbb{H}_0) = \mathbb{K}(G/H)$ with completion $L^2(G/H,\mathbb{H}_0;\mu) = L^2(G/H;\mu)$ under the scalar product

$$(\phi,\psi) \rightsquigarrow \int_{G/H} \phi(x)\bar{\psi}(x)d\mu(\dot{x}).$$

When $H = \{1_G\}$ we are back to the left regular representation $(\gamma_G, L^2(G))$ of G (see Section 1.22).

2.10 The unitary inducing procedure as described in the preceding section admits a "geometric" interpretation in the scenario of homogeneous vector bundles. Let G be a unimodular, locally compact, topological group and H a unimodular closed subgroup of G such there exists a continuous section of G fibred by H. Let $(U_0,\mathbb{H}_0)$ be a continuous, unitary, linear representation of H again. Then we may define a continuous *right* action of H on the topological product space $G \times \mathbb{H}_0$ as follows:

$$H \times (G \times \mathbb{H}_0) \ni (h,(x,f)) \rightsquigarrow (x,f).h = (xh, U_0(h^{-1})f) \in G \times \mathbb{H}_0.$$

The space E of all orbits (x,f).H equipped with the quotient topology is a Hausdorff topological space. The surjection $\pi:E \to G/H$ given by

$$\pi:E \ni (x,f).H \rightsquigarrow \dot{x} = \delta_G(H)x = x.H \in G/H$$

is continuous and open. Therefore the pair (E,π) is a *vector bundle* over the homogeneous space G/H. It is possible to endow (E,π) with the structure of a complex *Hilbert bundle over* G/H. Obviously there exists a bijection between the cross-sections s for (E,π) and the mappings $\phi_s:G \to \mathcal{H}_0$ satisfying with respect to the natural *right* H-action on G the covariance condition. The bijection $s \rightsquigarrow \phi_s$ is given via the prescription

$$s(\dot{x}) = (x,\phi_s(x)).H$$

for $x \in G$ and $\dot{x} = \delta_G(H)x = x.H \in G/H$. It follows that a cross-section s for (E,π) is continuous on the homogeneous space G/H if and only if $\phi_s \in \mathfrak{C}(G/H;\mathcal{H}_0)$. Let $d\mu = \frac{dx}{dh}$ denote a positive measure in $\mathfrak{M}(G/H)$ that is $\neq 0$ and invariant under the natural left action $(x,y) \rightsquigarrow \gamma_G(x)\dot{y}$ of G on G/H and form the complex Hilbert space $L^2(E;\mu)$ of quadratic integrable cross-sections for (E,π) with respect to μ. If we identify the cross-sections s,t with their associated elements ϕ_s, ψ_t in the completion $\mathcal{H}$ of $\mathcal{K}(G/H,\mathcal{H}_0)$, then

$$\langle s|t\rangle = \int_{G/H} \langle\phi_s(x)|\psi_t(x)\rangle_0 d\mu(\dot{x}) = \langle\phi_s|\psi_t\rangle_\mu .$$

Observe that G acts continuously on the *left* on (E,π) via

$$(x,(y,\phi).H) \rightsquigarrow \gamma_G(x)((y,\phi).H) = (\gamma_G(x)y,\phi).H \in E.$$

Consequently (E,π) forms a G-homogeneous Hilbert bundle and the bundle-theoretic definition of the unitarily induced representation $(U,\mathcal{H}) = \mathrm{Ind}_H^G(U_0,\mathcal{H}_0)$ reads as follows:

$$U(x)s = (\gamma_G(x)s) \circ \gamma_{G/H}(x^{-1})$$

provided $x \in G$ and the mapping ϕ_s associated to s belongs to the vector space $\mathcal{K}(G/H;\mathcal{H}_0)$.

2.11 Let G denote a unimodular, locally compact, topological group, H_1 a unimodular closed subgroup of G such that there exists a continuous section of G fibred by H_1, and $(U_1,\mathcal{H}_1)$ a continuous, unitary, linear representation of H_1 in the complex Hilbert space $\mathcal{H}_1$. Suppose that H_2 is a subgroup of G and the *conjugate* of H_1 by an element $x_0 \in G$. Thus the *inner automorphism*

$$\mathrm{Int}_G(x_0) : G \ni y \rightsquigarrow x_0 y\, x_0^{-1} \in G$$

of G *induced by* x_0 forms a topological isomorphism of H_1 onto H_2:

$$H_2 = \mathrm{Int}_G(x_0)H_1.$$

Suppose that $(U_2,\mathcal{H}_2)$ is a continuous, unitary, linear representation of H_2 in the complex Hilbert space $\mathcal{H}_2$ which is unitarily isomorphic to the continuous, unitary, linear representation

$$(x_0.U_1,\mathcal{H}_1) := (U_1 \circ \mathrm{Int}_G(x_0^{-1}),\mathcal{H}_1)$$

of H_2 in $\mathcal{H}_1$. If $T : \mathcal{H}_1 \to \mathcal{H}_2$ denotes a unitary isomorphism of $(x_0.U_1,\mathcal{H}_1)$ onto $(U_2,\mathcal{H}_2)$, the following diagram is commutative for all elements $h_1 \in H_1$:

$$\begin{array}{ccc} \mathcal{H}_1 & \xrightarrow{\ \ T\ \ } & \mathcal{H}_2 \\ {\scriptstyle U_1(x_o^{-1}\,h_2x_o)}\uparrow & & \uparrow{\scriptstyle U_2(h_2)} \\ \mathcal{H}_1 & \xrightarrow[\ \ T\ \ ]{} & \mathcal{H}_2 \end{array}$$

It can be easily checked that the mapping

$$G/H_1 \ni x.H_1 \rightsquigarrow (x\, x_0^{-1}).H_2 \in G/H_2$$

is well defined and a *homeomorphism* of the homogeneous space G/H_1 of left cosets of H_1 in G onto the homogeneous space G/H_2 of left cosets of H_2 in G. If $\phi_1 \in \mathcal{K}(G/H_1,\mathcal{H}_1)$ then

$$S\phi_1 : G \ni x \rightsquigarrow T(\phi_1(\,xx_0\,)) \in \mathcal{H}_2$$

is a continuous map of G into $\mathcal{H}_2$ which satisfies the covariance condition with respect to the natural right action $(h_2,x) \rightsquigarrow \delta_G(h_2)x$ of H_2 in G. Indeed, we have $h_2 = \mathrm{Int}_G(x_0)h_1 = x_0 h_1 x_0^{-1}$ for an element $h_1 \in H_1$ and so

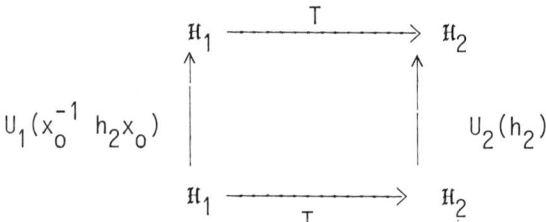

$$\delta_G(h_2)(S\phi_1(x)) = S\phi_1(xh_2) = s\phi_1(xx_0h_1x_0^{-1}) = T(\phi_1(xx_0h_1))$$

$$= T \circ \delta_G(h_1)(\phi_1(xx_0)) = T \circ U_1(h_1^{-1})(\phi_1(xx_0))$$

$$= (T \circ U_1(h_1) \circ T^{-1})^{-1} \circ T(\phi_1(xx_0))$$

$$= U_2(h_2^{-1}) \circ T(\phi_1(xx_0))$$

$$= U_2(h_2^{-1})(S\phi_1(x)).$$

Obviously $\mathrm{Supp}(\phi_1).x_0^{-1}$ is the compact support of $S\phi_1$ in G/H_2. Thus $S\phi_1 \in \mathbb{K}(G/H_2,\mathcal{H}_2)$ for every mapping $\phi_1 \in \mathbb{K}(G/H_1,\mathcal{H}_1)$. It follows from the essentially symmetric roles played by $\mathcal{H}_1$ and $\mathcal{H}_2$ in the preceding reasoning that the mapping

$$S : \mathbb{K}(G/H_1,\mathcal{H}_1) \to \mathbb{K}(G/H_2,\mathcal{H}_2)$$

is *surjective*. Choose Haar measures dx, dh_1, and dh_2 on G, H_1, and H_2, respectively, and form the quotient measures

$$d\mu_1 = \frac{dx}{dh_1} , \quad d\mu_2 = \frac{dx}{dh_2}$$

on the homogeneous spaces G/H_1, and G/H_2, respectively. It follows that S maps the separated complex prehilbert space $\mathbb{K}(G/H_1,\mathcal{H}_1; \langle\cdot|\cdot\rangle_{\mu_1})$ *isometrically* onto the separated complex prehilbert space $\mathbb{K}(G/H_2,\mathcal{H}_2; \langle\cdot|\cdot\rangle_{\mu_2})$. Therefore it admits a unique unitary, linear extension

$$S : L^2(G/H_1, \mathcal{H}_1;\mu_1) \to L^2(G/H_2,\mathcal{H}_2;\mu_2)$$

which satisfies the condition

$$S \circ \mathrm{Ind}_{H_1}^G U_1(x)\phi_1 = S \circ \gamma_{G/H_1}(x)\phi_1 = \mathrm{Ind}_{H_2}^G U_2(x) \circ S\phi_1$$

for any mapping $\phi_1 \in \mathbb{K}(G/H_1,\mathcal{H}_1)$ and all elements $x \in G$. Thus $S \in \mathcal{R}_G(\mathrm{Ind}_{H_1}^G U_1, \mathrm{Ind}_{H_2}^G U_2)$ and we established the following result:

2.12 <u>Theorem</u>. Let G be a unimodular, locally compact, topological group, H_1 a unimodular closed subgroup of G such that there exists a continuous section of G fibred by H_1, and $(U_1, \mathcal{H}_1)$ a continuous, unitary, linear representation of H_1. Let $H_2 = \mathrm{Int}_G(x_0)H_1$ be the subgroup of G conjugate of H_1 by $x_0 \in G$, and $(U_2, \mathcal{H}_2)$ a continuous, unitary, linear representation of H_2 which is unitarily isomorphic to $(x_0.U_1, \mathcal{H}_1)$. Then the unitarily induced linear representations $\mathrm{Ind}_H^G(H_\alpha, \mathcal{H}_\alpha)$ of G are unitarily isomorphic ($\alpha \in \{1,2\}$).

Corollary. For all elements $x_0 \in G$ the linear representations

$$\mathrm{Ind}_{H_1}^G (U_1, \mathcal{H}_1)$$

and

$$\mathrm{Ind}^G_{x_0 H x_0^{-1}} (x_0.U_1, \mathcal{H}_1)$$

are unitarily isomorphic.

2.13 Let us return to the situation considered in Section 2.6 and retain the notations used in this section. Moreover, suppose that the locally compact topological group G and its closed subgroups K, H are unimodular and that a continuous, unitary, linear representation $(U_0, \mathcal{H}_0)$ of K is given. Then we may form the unitarily induced linear representations

$$\mathrm{Ind}_K^G(U_0, \mathcal{H}_0)$$

and

$$\mathrm{Ind}_H^G(\mathrm{Ind}_K^H(U_0, \mathcal{H}_0))$$

of G. In order to show that these continuous, unitary, linear representations of G are unitarily isomorphic, set

$$(U, \mathcal{H}) = \mathrm{Ind}_K^G(U_0, \mathcal{H}_0),$$

$$(U_1, \mathcal{H}_1) = \mathrm{Ind}_K^H(U_0, \mathcal{H}_0),$$

$$(U_2, \mathcal{H}_2) = \mathrm{Ind}_H^G(U_1, \mathcal{H}_1).$$

For every $\phi \in \mathcal{K}(G/K, \mathcal{H}_0)$ and $x \in G$ define the mapping

$$T\phi(x) : H \ni h \rightsquigarrow \delta_G(h)\phi(x) \in \mathcal{H}_0.$$

If $x \in G$ is fixed and $k \in K$, then

$$\delta_G(k)T\phi(x) = U_o(k^{-1})T\phi(x),$$

hence $T\phi(x) \in \mathbb{K}(H/K,\mathcal{H}_o) \hookrightarrow \mathcal{H}_1$. The next step is to show that the mapping

$$T\phi : G \ni x \rightsquigarrow T\phi(x) \in \mathcal{H}_1$$

is continuous. Let $x_o \in G$ be a fixed element. For every element $x \in G$ the function

$$\phi_x : H \ni h \rightsquigarrow \|\phi(xh) - \phi(x_o h)\|_o^2 \in \mathbb{R}_+$$

belongs to the vector space $\mathbb{K}(H/K)$. Let V be a compact neighbourhood of 1_G in G. There exists a compact subset M of H/K such that

$$\bigcup_{x \in V.x_o} \mathrm{Supp}(\phi_x) \subseteq M.$$

Choose a compact subset L of H such that $q_o(L) = M$ holds. Since the function

$$V.x_o \times L \ni (x,h) \rightsquigarrow \|\phi(xh) - \phi(x_o h)\|_o^2 \in \mathbb{R}_+$$

is uniformly continuous, there exists for any $\varepsilon > 0$ a neighbourhood W of 1_G in G such that

$$\|\phi(xh) - \phi(x_o h)\|_o^2 < \frac{\varepsilon}{\alpha(M)}$$

for all pairs $(x,h) \in W.x_o \times L$. If $\|\cdot\|_1$ denotes the norm of the complex Hilbert space $\mathcal{H}_1$ then we have

$$\|T\phi(x) - T\phi(x_o)\|_1^2 = \int_{H/K} \|\phi(xh) - \phi(x_o h)\|_o^2 \, d\alpha(h.K) \leqq \varepsilon$$

for $x \in W.x_o$. This proves $T\phi \in \mathfrak{C}(G,\mathcal{H}_1)$. In view of the identities

$$\delta_G(h)T\phi = U_1(h^{-1})T\phi$$

for $h \in H$, and

$$\mathrm{Supp}(T\phi) = \mathrm{Supp}(\phi).H$$

it follows $T\phi \in \mathbb{K}(G/H,\mathcal{H}_1)$. The linear mapping

$$\mathcal{K}(G/K,\mathcal{H}_0) \ni \phi \rightsquigarrow T\phi \in \mathcal{K}(G/H,\mathcal{H}_1)$$

is isometric. Indeed, if $\phi \in \mathcal{K}(G/K,\mathcal{H}_0)$ then the transivity of invariant measures established in Theorem 2.7 yields

$$\begin{aligned}
||T\phi||^2 &= \int_{G/H} ||T\phi(x)||_1^2 \, d\lambda(\dot{x}) \\
&= \int_{G/H} \Big(\int_{H/K} ||\delta_G(h)\phi(x)||_0^2 \, d\alpha(hK)\Big) d\lambda(\dot{x}) \\
&= \int_{G/K} ||\phi(y)||_0^2 \, d\beta(yK) \\
&= ||\phi||^2.
\end{aligned}$$

Let $T : \mathcal{H} \to \mathcal{H}_2$ also denote the unique isometric extension of the linear mapping $\phi \rightsquigarrow T\phi$. Clearly the intertwining identity

$$T \circ U_2(x) = U(x) \circ T$$

holds for all elements $x \in G$. For $\psi \in \mathcal{K}(G/H,\mathcal{H}_1)$ define the mapping

$$S\psi : G \ni x \rightsquigarrow (H \ni h \rightsquigarrow \delta_G(1_G)\psi(x) \in \mathcal{H}_1) \in \mathcal{K}(G/H,\mathcal{H}_0).$$

Then $S \in \mathcal{K}(G/K,\mathcal{H}_0)$ and $||S\psi|| = ||\psi||$. Let $S : \mathcal{H}_2 \to \mathcal{H}$ also denote the unique isometric extension of the linear mapping $\psi \rightsquigarrow S\psi$. Clearly $S \circ T = id_{\mathcal{H}}$ and $T \circ S = id_{\mathcal{H}_2}$. Thus we established the following result:

2.14 Theorem. Let G denote a unimodular, locally compact, topological group and let H be a closed subgroup of G so that H is unimodular and such that there exists a continuous section of G fibred by H. Let K be a closed subgroup of H so that K is unimodular and such that there exists a continuous section of G fibred by K. If $(U_0,\mathcal{H}_0)$ denotes a continuous, unitary, linear representation of K in the complex Hilbert space $\mathcal{H}_0$, then the unitarily induced linear representations $\mathrm{Ind}_K^G(U_0,\mathcal{H}_0)$ and $\mathrm{Ind}_H^G(\mathrm{Ind}_K^H(U_0,\mathcal{H}_0))$ are unitarily isomorphic.

The preceding result on unitarily inducing the linear representation $(U_0,\mathcal{H}_0)$ of K in '*stages*' will sometimes be written in the form

$$\mathrm{Ind}_H^G(\mathrm{Ind}_K^H(U_0,\mathcal{H}_0)) = \mathrm{Ind}_K^G(U_0,\mathcal{H}_0)$$

because the unitary isomorphism T between the continuous, unitary, linear representations of G on the left and the right hand side is defined quite naturally.

2.15 Let G' denote a *closed normal* subgroup of the locally compact topological group G. Form the locally compact quotient group G" = G/G' and the continuous canonical epimorphism $\pi : G \to G"$. Suppose that there exists a continuous section of G fibred by G'. Let H" be a *closed* subgroup of G" such that there exists a continuous section of G" fibred by H" and the homogeneous space G"/H" admits a positive measure $\mu" \neq 0$ that is invariant under the natural left action of G" on G"/H". If $H = \pi^{-1}(H")$ is the preimage of H" in G under π and $\pi_0 = \pi | H$, the following diagram may illustrate the present situation:

$$\begin{array}{ccccc} G' & \hookrightarrow & H & \hookrightarrow & G \\ \downarrow & & \downarrow \pi_0 & & \downarrow \pi \\ \{1\}_{G"} & \hookrightarrow & H" & \hookrightarrow & G" \end{array}$$

Suppose that a continuous, unitary, linear representation $(U"_0, \mathcal{H}"_0)$ of H" is given. Form the unitarily induced representation $(U", \mathcal{H}") = \mathrm{Ind}_{H"}^{G"}(U"_0, \mathcal{H}"_0)$ of G". The following technical lemma shows that the unitary inducing procedure is compatible with the natural pull-backs.

2.16 <u>Lemma</u>. In the notations and under the assumptions of Section 2.13 the unitarily induced representations $(U" \circ \pi, \mathcal{H}")$ and $\mathrm{Ind}_H^G(U"_0 \circ \pi_0, \mathcal{H}"_0)$ of G are unitarily isomorphic

<u>Proof</u>. Consider the canonical continuous surjections $p : G \to G/H$ and $p" : G" \to G"/H"$. There exists a unique homeomorphism $\pi"$ of G/H onto G"/H" which makes the following diagram commutative:

$$\begin{array}{ccc} G & \xrightarrow{\ p\ } & G/H \\ \pi \downarrow & & \downarrow \pi" \\ G" & \xrightarrow[\ p"\ ]{} & G"/H" \end{array}$$

For any mapping $\phi \in \mathcal{K}(G/H,\mathcal{H}_0'')$ we have the covariance condition

$$\delta_G(h)\phi = U_0''(\pi_0(h)^{-1})\phi$$

for all elements $h \in H$. In particular it follows

$$\delta_G(x')\phi = \phi$$

for all $x' \in G'$. Therefore ϕ determines a unique continuous mapping $\Phi: G'' \to \mathcal{H}_0''$ such that

$$\phi = \Phi \circ \pi$$

holds. Obviously the identities

$$\delta_{G''}(h'')\Phi = U_0''(h''^{-1})\Phi$$

for all $h'' \in H''$ and

$$p(\mathrm{Supp}(\phi)) = \pi''^{-1}(p''(\mathrm{Supp}(\Phi)))$$

imply $\Phi \in \mathcal{K}(G''/H'',\mathcal{H}_0'')$. Thus the linear mapping T defined via the prescription

$$T: (G/H,\mathcal{H}_0'') \ni \phi \quad > \Phi \in \mathcal{K}(G''/H'',\mathcal{H}_0'')$$

is a bijection. Since $\mu'' \in \mathfrak{M}(G''/H'')$ is invariant under the natural left action of G'' on G''/H'', its preimage $\mu = \pi''^{-1}(\mu'') \in \mathfrak{M}(G/H)$ is invariant under the natural left action of G on the homogeneous space G/H and we have

$$\int_{G/H} ||\phi(x)||^2 \, d\mu(\dot{x}) = \int_{G''/H''} ||\Phi(x'')||^2 \, d\mu''(\dot{x}'').$$

Therefore T defines an isometric linear bijection of $\mathcal{K}(G/H,\mathcal{H}_0'')$ onto $\mathcal{K}(G''/H'',\mathcal{H}_0'')$. Observe that there exists a continuous section of G fibred by H. The unique unitary linear extension of T to the representation space $L^2(G/H,\mathcal{H}_0'';\mu)$ of $\mathrm{Ind}_H^G(U_0'' \circ \pi_0,\mathcal{H}_0'')$ onto the representation space $L^2(G''/H'',\mathcal{H}_0'';\mu'')$ of $\mathrm{Ind}_{H''}^{G''}(U_0'',\mathcal{H}_0'')$ is an isomorphism of $\mathrm{Ind}_H^G(U_0'' \circ \pi_0,\mathcal{H}_0'')$ onto $(U'' \circ \pi, \mathcal{H}'')$. ——

2.17 It is a surprising fact that to some extent the known infinite-dimensional, topologically irreducible, continuous, unitary, linear representations of locally compact topological groups are monomial (cf. 2.9,

Remark 3). A locally compact topological group G is said to be *monomial* if a complete set of representatives of its unitary dual can be constructed by unitarily inducing one-dimensional, continuous, linear representations, i.e., by unitarily inducing continuous unitary characters of suitably chosen closed subgroups of G. In order to examine situations where such a procedure is possible, suppose that there exists a *closed normal abelian subgroup* N sitting inside the locally compact topological group G. Let $\hat{N}$ denote the Pontryagin dual of N. Then $\hat{N}$ carries the topology of uniform convergence on the compact subsets of N under which it is a locally compact, abelian, topological group. Since N is stable for all $x \in G$ under the action of the inner automorphism of N induced by x (cf. Section 2.11), the mapping

$$G \times \hat{N} \ni (x,\hat{y}) \rightsquigarrow x.\hat{y} = \hat{y} \circ \mathrm{Int}_N(x^{-1}) \in \hat{N}$$

defines a *continuous left action* of the locally compact group G on the locally compact, abelian group $\hat{N}$. We shall call it the *left action of* G *on* $\hat{N}$ *by inner automorphisms of* N. From the Corollary of Theorem 2.12 supra we conclude

2.18 Theorem. Let N be a closed, normal, abelian subgroup of the locally compact topological group G such that there exists a continuous section of G fibred by N. Let $\hat{x} \in \hat{N}$, $\hat{y} \in \hat{N}$ denote two continuous unitary characters of N which belong to the same orbit of the left action of G on $\hat{N}$ by inner automorphisms of N, i.e., $\hat{x} \in G.\hat{y}$. Then the monomial representations

$$\mathrm{Ind}_N^G(\hat{x}.\mathrm{id}_{\mathbb{C}},\mathbb{C}), \quad \mathrm{Ind}_N^G(\hat{y}.\mathrm{id}_{\mathbb{C}},\mathbb{C})$$

of G are unitarily isomorphic.

Thus in order to study within isomorphy the monomial representations unitarily induced by continuous unitary characters $\hat{y}$ of N it will be sufficient to induce unitarily a *generic* element $\hat{x} \in \hat{N}$ of the orbit $G.\hat{y}$ of the left action of G on $\hat{N}$ by inner automorphisms of N.

The next lemma includes a technical result which holds for any closed, normal, abelian subgroup N of an arbitrary locally compact, topological group G.

2.19 Lemma. Let $\hat{y} \in \hat{N}$ denote a continuous unitary character of N such that its orbit $G.\hat{y}$ under the left action of G on $\hat{N}$ by inner automorphisms of N is a locally compact topological subspace of the locally compact topological group $\hat{N}$. Then the continuous mapping

$$\sigma_{\hat{y}} : G \ni x \rightsquigarrow x.\hat{y} \in G.\hat{y}$$

associated with $\hat{y} \in \hat{N}$ is an open map.

Proof. Let V denote an arbitrary neighbourhood of the element 1_G in G. Choose a compact neighbourhood V_o of 1_G in G such that $V_o^{-1} = V_o$ and $V_o^2 \subseteqq V$. Since G is countable at infinity there exists a sequence $(x_n)_{n\in\mathbb{N}}$ of elements of G such that we have

$$G = \bigcup_{n \in \mathbb{N}} (x_n.V_o).$$

It follows that the orbits of the continuous unitary characters $\hat{y} \in \hat{N}$ of N satisfy

$$G.\hat{y} = \bigcup_{n \in \mathbb{N}} (x_n.V_o.\hat{y}).$$

Since $G.\hat{y}$ is a Baire space there exists a number $n_o \in \mathbb{N}$ such that the compact subset $x_{n_o}.V_o\hat{y}$ of $G.\hat{y}$ has a non-void interior. It follows that $V_o.\hat{y}$ admits an interior point and therefore that $\hat{y}$ is an interior point of $V_o^2.\hat{y} \subseteqq V.\hat{y}$. Consequently the image $\sigma_{\hat{y}}(V)$ of V under $\sigma_{\hat{y}}$ is a neighbourhood of $\hat{y}$ in $G.\hat{y}$, whence the lemma. —

A closed, normal, abelian subgroup N of a locally compact topological group G is said to be *properly embedded into* G, provided the orbit $G.\hat{y}$ of each unitary character $\hat{y} \in \hat{N}$ under the left action of G on $\hat{N}$ by inner automorphisms of N is a locally compact topological subspace of the locally compact topological group $\hat{N}$. Let $G_{\hat{y}}$ denote the *stabilizer* of $\hat{y} \in \hat{N}$ in G which is, of course, a closed subgroup of G containing N. Thus we have the ascending filtration

$$N \hookrightarrow G_{\hat{y}} \hookrightarrow G \qquad (\hat{y} \in \hat{N}).$$

In view of the preceding lemma the natural bijection

$$G/G_{\hat{y}} \to G.\hat{y}$$

obtained by factorizing canonically the mapping $\sigma_{\hat{y}}$ is a *homeomorphism* for each $\hat{y} \in \hat{N}$ if and only if N is properly embedded into G. In this case, $\sigma_{\hat{y}}$ allows to identify the locally compact topological spaces $G/G_{\hat{y}}$ and $G.\hat{y}$ and to state the following version of the *Mackey Little Group Theorem.*

2.20 Theorem (Mackey). Let G denote a unimodular, locally compact, topological group and N a closed, normal, abelian subgroup which is properly embedded into G.

(i) For every topologically irreducible, continuous, unitary, linear representation $(U,\mathcal{H})$ of G there exists one and only one orbit $\mathcal{O}$ of the left action of G on $\hat{N}$ by inner automorphisms of N with the following property: There exists for every unitary character $\hat{x} \in \mathcal{O}$ of N a continuous, unitary, linear representation $(U_0,\mathcal{H}_0)$ of its stabilizer $G_{\hat{x}}$ in G such that

$$U_0(x) = \hat{x}(x).id_{\mathcal{H}_0}$$

holds for all elements $x \in N$ and that, moreover, the unitarily induced linear representation

$$Ind_{G_{\hat{x}}}^{G}(U_0,\mathcal{H}_0)$$

of G is isomorphic to $(U,\mathcal{H})$.

(ii) For every orbit $\mathcal{O}$ of G in $\hat{N}$ and every element $\hat{x} \in \mathcal{O}$ the mapping

$$(U_0,\mathcal{H}_0) \rightsquigarrow Ind_{G_{\hat{x}}}^{G}(U_0,\mathcal{H}_0)$$

of the family of those topologically irreducible, continuous, unitary, linear representations $(U_0,\mathcal{H}_0)$ of $G_{\hat{x}}$ such that $U_0(x) = \hat{x}(x).id_{\mathcal{H}_0}$ $(x \in N)$ onto the family of topologically irreducible, continuous, unitary, linear representations of G is bijective and factors through the equivalence of being unitarily isomorphic.

The proof of this result and the next one are omitted. Both results form parts of the so-called *Mackey machinery* which is centered around the imprimitivity theorem.

Let us agree to retain the above notations and the assumptions about G and N as in the preceding theorem. Assume in addition that G is the *semi-direct product of* N *with a closed subgroup* K *of* G. Since N is abelian the left actions of G on N and $\hat{N}$, respectively, by inner automorphisms of N are equivalent to the left actions of K on N and $\hat{N}$, respectively, by inner automorphisms of N. Choose an element $\hat{x}$ in an orbit $\mathcal{O}$ of the left action of G on $\hat{N}$. For each topologically irreducible, continuous, unitary, linear representation $(U_0, \mathcal{H}_0)$ of the closed subgroup $G_{\hat{x}} \cap K$ of G set

$$(\hat{x} \times U_0)(kx) = \hat{x}(x).U_0(k)$$

where $k \in G_{\hat{x}} \cap K$ and $x \in K$. This prescription defines a topologically irreducible, continuous, unitary, linear representation of $G_{\hat{x}}$.

Corollary. Suppose that G is the semi-direct product of the properly embedded, closed, normal, abelian subgroup N and the closed subgroup K. Let $(U, \mathcal{H})$ denote a continuous, unitary, linear representation of G - then there exists an orbit $\mathcal{O}$ of the left action of G on $\hat{N}$ by inner automorphisms of N such that for any unitary character $\hat{x} \in \mathcal{O}$ of N and a suitable continuous, unitary, linear representation $(U_0, \mathcal{H}_0)$ of $G_{\hat{x}} \cap K$ the unitarily induced linear representation

$$\mathrm{Ind}_{G_x}^{G} (\hat{x} \times U_0, \mathcal{H}_0)$$

of G is isomorphic to $(U, \mathcal{H})$.

Remark. In the case when G is a connected, simply connected, nilpotent, real Lie group with one-dimensional centre C, there exists a properly embedded closed normal abelian subgroup N of G. Moreover, if N is connected then there exists a continuous section of G fibred by N. The left action of G on $\hat{N}$ by inner automorphisms of N leads in a natural way to the coadjoint action of G in the dual vector space of the (abelian) Lie algebra of N; see 4.17 infra. It follows that there is a link between the Mackey orbital analysis on the one hand and the Kirillov coadjoint orbit theory for simply connected nilpotent Lie groups on the other hand.

References

Mackey, G.W. : Induced representations of groups and quantum mechanics. W.A. Benjamin, New York, Amsterdam, and Editore Boringhieri, Torino 1968.

Mackey, G.W. : Induced representations of locally compact groups and applications. In: Functional analysis and related fields, pp. 132-166. Edited by F.E. Browder. Springer, Berlin, Heidelberg, New York 1970.

Mackey, G.W. : Unitary group representations in physics, probability, and number theory. Benjamin/Cummings, Reading, Massachusetts 1978.

Warner, G. : Harmonic analysis on semi-simple Lie groups I. Die Grundlehren der mathematischen Wissenschaften, Band 188. Springer, Berlin, Heidelberg, New York 1972.

3 Square integrable linear group representations

3.1 Let G denote a unimodular, locally compact, topological group with left (hence right) Haar measure dx and $(U_0,\mathcal{H}_0)$ a continuous, unitary, linear representation of G in the complex Hilbert space $\mathcal{H}_0$. Recall (Section 1.14) that for each pair $(f,g) \in \mathcal{H}_0 \times \mathcal{H}_0$ the coefficient $c_{U_0,f,g}$ of $(U_0,\mathcal{H}_0)$ belongs to the vector space $\mathfrak{C}^b(G)$ of bounded, continuous, complex-valued functions on G. If G is a real Lie group and $(f,g) \in \mathcal{H}_0^\infty \times \mathcal{H}_0^\infty$ (cf. 1.2, Remark 3), then $c_{U_0,f,g} \in \mathfrak{C}^\infty(G)$.

The continuous, unitary, linear representation $(U_0,\mathcal{H}_0)$ of the unimodular, locally compact, topological group G is said to be *square integrable*, if $(U_0,\mathcal{H}_0)$ is *topologically irreducible* and the coefficients $c_{U_0,f,g}$ of $(U_0,\mathcal{H}_0)$ *belong to the complex Hilbert space* $L^2(G) = L^2(G;dx)$ for *all* pairs $(f,g) \in \mathcal{H}_0 \times \mathcal{H}_0$. In this case the linear representation $(U_0,\mathcal{H}_0)$ is said to belong to the *discrete series* (of topologically irreducible, continuous, unitary, linear representations) of G.

Remarks.

1. It can be proved by means of the topological version of Schur's theorem (Theorem 1.9) that for every topologically irreducible, continuous, unitary, linear representation $(U_0,\mathcal{H}_0)$ of G the existence of some *non-zero* vectors f_0,g_0 in $\mathcal{H}_0$ such that $c_{U_0,f_0,g_0} \in L^2(G)$ implies $c_{U_0,f,g} \in L^2(G)$ for *all* pairs $(f,g) \in \mathcal{H}_0 \times \mathcal{H}_0$; cf. Theorem 3.3 infra. Thus the hypotheses of square integrability of $(U_0,\mathcal{H}_0)$ can be considerably weakened. The notion can be generalized to cover non-unimodular locally compact groups, but this is of no importance here.

2. The topologically irreducible, continuous, unitary, linear representations $(U_0,\mathcal{H}_0)$ of any *compact* topological group G are square integrable.

The existence of a square integrable linear representation $(U_0,\mathcal{H}_0)$ has an immediate implication on the structure of G.

3.2 Lemma. Suppose that the unimodular, locally compact, topological group G admits a square integrable linear representation (U_0, H_0) - then the centre C of G is compact.

Proof. Let χ_{U_0} denote the central character (1.6) of (U_0, H_0) and consider the one-dimensional continuous, unitary, linear representation $(\chi_{U_0}.\mathrm{id}_{\mathbb{C}}, \mathbb{C})$ of the closed subgroup C of G. For every pair of vectors $(f,g) \in H_0 \times H_0$ the coefficient $c_{U_0,f,g}$ of (U_0, H_0) satisfies with respect to the natural right (hence left) action $(z,x) \rightsquigarrow \delta_G(z)x$ of C on G the covariance condition

$$\delta_G(z)c_{U_0,f,g} = \chi_{U_0}(z)c_{U_0,f,g}$$

for all elements $z \in C$. It follows that the functions $c_{U_0,f,g}$ are constant on the left (hence right) cosets $\dot{x} = \delta_G(C)x = x.C$ modulo C. Choose on the unimodular, locally compact, topological groups C and G/C Haar measures dz and $d\mu$, respectively, such that $d\mu = \frac{dx}{dz}$ holds - then we have the identity (cf. Section 2.3)

$$\int_G |c_{U_0,f,g}(x)|^2 dx = \left(\int_{G/C} |c_{U_0,f,g}|^2(\dot{x})d\mu(\dot{x})\right).\left(\int_C dz\right)$$

for all pairs $(f,g) \in H_0 \times H_0$. Thus $c_{U_0,f_0,g_0} \in L^2(G)$ for any two non-zero vectors f_0, g_0 in H_0 implies that C has finite Haar measure. Consequently the central normal subgroup C of G is compact. ——

The preceding lemma suggests the following extension of the notion of square integrable linear representation. Let Z denote a *closed* subgroup of the centre C of the unimodular, locally compact, topological group G - then we have for each topologically irreducible, continuous, unitary, linear representation (U_0, H_0) of G the inclusions

$$Z \hookrightarrow C \hookrightarrow \mathfrak{P}_G(U_0)$$

where $\mathfrak{P}_G(U_0)$ denotes the projective kernel of (U_0, H_0) in G (1.6). Suppose that Z is *cocompact* in C, i.e., that the quotient C/Z is a *compact* group. Choose Haar measures dz and $d\mu$ on the unimodular, locally compact, topological groups Z and G/Z, respectively, such that $d\mu = \frac{dx}{dz}$ holds. The continuous, unitary, linear representation (U_0, H_0) of G is said to be *square integrable modulo* Z if (U_0, H_0) is topologically irreducible and the continuous bounded functions

$$G/Z \ni \dot{x} \rightsquigarrow |c_{U_0,f,g}|(\dot{x}) \in \mathbb{R}_+$$

belong to the Hilbert space $L^2(G/Z;\mu)$ for all pairs $(f,g) \in \mathcal{H}_0 \times \mathcal{H}_0$ (note that the absolute value of $c_{U_0,f,g}$ is constant on cosets modulo $\mathcal{P}_G(U_0)$). If the centre C is compact in G, the notion of square integrability modulo C is, of course, equivalent to the notion of square integrability as defined in Section 3.1.

3.3 <u>Theorem</u>. Let $(U_0,\mathcal{H}_0)$ denote a topologically irreducible, continuous, unitary, linear representation of the unimodular, locally compact, topological group G with central character χ_{U_0}. Let Z be a closed cocompact subgroup of the centre C of G and $\chi = \chi_{U_0}|Z$ the central character of $(U_0,\mathcal{H}_0)$ on Z - then the following three conditions are mutually equivalent:

(i) $(U_0,\mathcal{H}_0)$ is square integrable modulo Z;

(ii) There exist two non-zero vectors f_0,g_0 in $\mathcal{H}_0$ such that $c_{U_0,f_0,g_0} \in L^2(G/Z;\mu)$;

(iii) $(U_0,\mathcal{H}_0)$ is contained in the monomial representation $\mathrm{Ind}_Z^G(\bar{\chi}\mathrm{id}_{\mathbb{C}},\mathbb{C})$ of G.

If $(U_0,\mathcal{H}_0)$ satisfies one of these conditions, then its coefficients $c_{U_0,f,g}$ belong to the representation space $\mathcal{H} = L^2(G/Z,\bar{\chi};\mu)$ of the monomial representation $\mathrm{Ind}_Z^G(\bar{\chi}.\mathrm{id}_{\mathbb{C}},\mathbb{C})$ for all pairs $(f,g) \in \mathcal{H}_0 \times \mathcal{H}_0$.

<u>Proof</u>. Recall that $\mathcal{H}$ is the completion of the complex prehilbert space $\mathcal{K}(G/Z,\bar{\chi})$ under the scalar product inherited from the complex Hilbert space $L^2(G/Z;\mu)$. In particular, $\mathcal{H}$ is stable under the left and right convolution by elements of the vector space $\mathfrak{M}^1(G)$ of bounded complex measures on G. The identities

$$c_{U_0,U_0^1(\nu)f,g} = c_{U_0,f,g} * \check{\nu},$$

and

$$c_{U_0,f,U_0^1(\nu)g} = \bar{\nu} * c_{U_0,f,g}$$

which hold for all pairs $(f,g) \in \mathcal{H}_0 \times \mathcal{H}_0$ and all measures $\nu \in \mathfrak{M}^1(G)$ show that for all vectors $(f_0,g_0) \in \mathcal{H}_0 \times \mathcal{H}_0$ satisfying the condition (ii) the sets given by

$$\mathcal{L}_{g_0} = \{f \in \mathcal{H}_0 | c_{U_0,f,g_0} \in \mathcal{H}\}$$

and

$$\mathcal{K}_{f_0} = \{g \in \mathcal{H}_0 | c_{U_0,f_0,g} \in \mathcal{H}\}$$

form everywhere dense vector subspaces of $\mathcal{H}$ that are stable under the linear, resp., antilinear action of $U_0^1(\mathfrak{m}^1(G))$. The $\mathbb{C}$-linear mapping S_{g_0} defined via the prescription

$$S_{g_0} : \mathcal{L}_{g_0} \ni f \rightsquigarrow c_{U_0,f,g_0} \in \mathcal{H}$$

has a closed graph and satisfies the identity

$$\begin{aligned} S_{g_0} \circ U(x)f &= c_{U_0,f,g_0} * \varepsilon_{x^{-1}} \\ &= \delta_{Z\setminus G}(x) \circ S_{g_0}(f) \end{aligned}$$

for all $f \in \mathcal{L}_{g_0}$ and all elements $x \in G$. An application of the topological version of Schur's Theorem (Theorem 1.9 supra) shows that S_{g_0} extends to a continuous $\mathbb{C}$-linear mapping $S_{g_0} : \mathcal{H}_0 \to \mathcal{H}$ that forms an isomorphism of $(U_0,\mathcal{H}_0)$ onto a linear subrepresentation of the continuous, unitary, linear representation $(\delta_{Z\setminus G},\mathcal{H})$. In particular, we have $c_{U_0,f,g_0} \in \mathcal{H}$ and $S_{g_0}(f) = c_{U_0,f,g_0}$ for all vectors $f \in \mathcal{H}_0$. An analogous reasoning based on the $\mathbb{C}$-linear mapping

$$T_{f_0} : \mathcal{K}_{f_0} \ni g \rightsquigarrow \bar{c}_{U_0,f_0,g} \in \mathcal{H}$$

shows that the condition (ii) implies (i) as well as (iii). The proof is complete if we can show that (iii) implies (ii).

Consider $\mathcal{H}_0$ as a closed vector subspace of $\mathcal{H}$ that is stable under the monomial representation $\mathrm{Ind}_Z^G(\bar{\chi}.\mathrm{id}_{\mathbb{C}},\mathbb{C})$ of G. Let $\mathcal{K}_0$ be the image of $\mathcal{K}(G/Z,\bar{\chi})$ under the orthoprojector of $\mathcal{H}$ onto $\mathcal{H}_0$. Since $\mathcal{K}(G/Z,\bar{\chi})$ is an everywhere dense vector subspace of $\mathcal{H}$, the vector subspace $\mathcal{K}_0$ is everywhere dense in $\mathcal{H}_0$. Let

$f_o \in \mathcal{H}_o$, $f_o \neq 0$, and $g_o \in \mathcal{K}_o$. Choose an element $g \in \mathcal{K}(G/Z,\bar{\chi})$ in the fibre over g_o - then we have

$$c_{U_o,f_o,g_o} = c_{U_o,f_o,g}.$$

In view of the identities

$$c_{U_o,f_o,g}(x) = \int_{G/Z} \gamma_{G/Z}(x) f_o(\dot{y})\bar{g}(\dot{y})d\mu(\dot{y})$$

$$= \int_{G/Z} \bar{g}(\dot{y})\check{f}_o(\dot{y}^{-1}x)d\mu(\dot{y})$$

$$= \bar{g} * f_o(x)$$

which hold for all elements $x \in G$, we obtain the following estimate:

$$\|c_{U_o,f_o,g_o}\|_2 \leq \|g\|_1 \cdot \|f_o\|_2$$

Thus $c_{U_o,f_o,g_o} \in \mathcal{H}$. —

3.4 Theorem (Frobenius-Schur-Godement). Let $(U_o,\mathcal{H}_o)$ denote a linear representation of the unimodular, locally compact, topological group which is square integrable modulo the centre C of G - then there exists a real number $\deg(U_o) > 0$ such that

$$\langle c_{U_o,f_o,g_o} | c_{U_o,f'_o,g'_o}\rangle = \frac{1}{\deg(U_o)} \langle f_o|f'_o\rangle . \langle g'_o|g_o\rangle$$

holds for all vectors f_o, f'_o, g_o, g'_o in $\mathcal{H}_o$ where the bracket $\langle\cdot|\cdot\rangle$ on the left hand side denotes the scalar product of the representation space $\mathcal{H} = L^2(G/C,\bar{\chi};\mu)$ of $\mathrm{Ind}_{\mathbb{C}}^{G}(\bar{\chi}_{U_o}.\mathrm{id}_{\mathbb{C}},\mathbb{C})$ and the brackets on the right hand side denote the scalar product of the complex Hilbert space $\mathcal{H}_o$.

Proof. It is known from the proof of Theorem 3.3 that for any non-zero vector $g_o \in \mathcal{H}_o$ the continuous $\mathbb{C}$-linear mapping

$$S_{g_o} : \mathcal{H}_o \ni f \rightsquigarrow c_{U,f,g_o} \in \mathcal{H}$$

forms an isomorphism of $(U_o,\mathcal{H}_o)$ onto a linear subrepresentation of $(\delta_{C\backslash G},\mathcal{H})$. Thus we have

$$S^*_{g'_0} \circ S_{g_0} \in \mathcal{R}_G(U_0)$$

for all pairs $(g_0, g'_0) \in \mathcal{H}_0 \times \mathcal{H}_0$. Corollary 1 of Theorem 1.6 supra then implies the existence of a constant $a_{g_0,g'_0} \in \mathbb{C}$ such that

$$\begin{aligned}
\langle c_{U_0,f_0,g_0} | c_{U_0,f'_0,g'_0} \rangle &= \langle S_{g_0}(f_0) | S_{g'_0}(f'_0) \rangle \\
&= \langle S^*_{g'_0} \circ S_{g_0}(f_0) | f'_0 \rangle \\
&= a_{g_0,g'_0} \langle f_0 | f'_o \rangle
\end{aligned}$$

holds for all pairs $(f_0, f'_0) \in \mathcal{H}_0 \times \mathcal{H}_0$. On the other hand we have the identities

$$\begin{aligned}
\langle c_{U_0,f_0,g_0} | c_{U_0,f'_0,g'_0} \rangle &= \int_{G/Z} \langle U_0(x) f_0 | g_0 \rangle \overline{\langle U_0(x) f'_0 | g'_0 \rangle} d\mu(\dot{x}) \\
&= \int_{G/Z} \langle f_0 | U_0(x^{-1}) g_0 \rangle \overline{\langle f'_0 | U_0(x^{-1}) g'_0 \rangle} d\mu(\dot{x}) \\
&= \int_{G/Z} \langle f_0 | U_0(x) g_0 \rangle \overline{\langle f'_0 | U_0(x) g'_0 \rangle} d\mu(\dot{x}) \\
&= \int_{G/Z} \langle U_0(x) g'_0 | f'_0 \rangle \overline{\langle U_0(x) g_0 | f_0 \rangle} d\mu(\dot{x}) \\
&= \langle c_{U_0,g'_0,f'_0} | c_{U_0,g_0,f_o} \rangle.
\end{aligned}$$

Hence

$$\langle c_{U_0,f_0,g_0} | c_{U_0,f'_0,g'_0} \rangle = a_{f'_0,f_0} \langle g'_0 | g_0 \rangle.$$

We conclude by comparison that the identity

$$a_{g_0,g'_0} \langle f_0 | f'_0 \rangle = a_{f'_0,f_0} \langle g'_0 | g_0 \rangle$$

holds for all vectors f_0, f'_0, g_0, g'_0 in $\mathcal{H}_0$, and therefore that the quotient

$$\frac{a_{f'_0,f_0}}{\langle f_0 | f'_0 \rangle}$$

is a real constant > 0 for all non-orthogonal elements $f_0 \neq 0$, $f'_0 \neq 0$ in $\mathcal{H}_0$. If we denote this constant by $\frac{1}{\deg(U_0)}$, the theorem follows. —

The number $\deg(U_0) > 0$ is called the *formal degree* of the square integrable linear representation $(U_0,\mathcal{H}_0)$ of G. It depends upon the choice of the Haar measure μ on G/C and merely upon the unitary isomorphy class of $(U_0,\mathcal{H}_0)$.

Corollary. The coefficients of $(U_0,\mathcal{H}_0)$ satisfy the convolution identity

$$c_{U_0,f_0,g_0} * c_{U_0,f'_0,g'_0} = \frac{1}{\deg(U_0)} \langle f_0 | g'_0 \rangle c_{U_0,f'_0,g_0}$$

for all vectors f_0, g_0, f'_0, g'_0 in $\mathcal{H}_0$.

Proof. Since we have

$$\langle c_{U_0,f_0,g_0} | c_{U_0,f'_0,g'_0} \rangle = c_{U_0,f_0,g_0} * c^{*}_{U_0,f'_0,g'_0}(1_G),$$

the identity to be proved follows from Theorem 3.4 by means of a translation.

Remark. In the special case when G denotes the real Heisenberg nilpotent Lie group $\tilde{A}(\mathbb{R})$ with one-dimensional centre $\tilde{C}$ (see Section 5.4 infra) and χ a non-trivial continuous character of C, the convolution structure of the representation space $L^2(G/\tilde{C},\chi;\mu \otimes \mu)$ of the monomial representation $\mathrm{Ind}_{\tilde{C}}^{G}(\chi.\mathrm{id}_{\mathbb{C}},\mathbb{C})$ of G plays an important role in harmonic analysis on G (μ = Lebesgue measure of $\mathbb{R}$); cf. Section 7 infra.

3.5 Theorem. Let $(U_1,\mathcal{H}_1)$ and $(U_2,\mathcal{H}_2)$ denote two continuous, unitary, linear representations of the unimodular, locally compact, topological group G. Suppose that $(U_1,\mathcal{H}_1)$ is topologically irreducible and that

$$U_2(z) = \chi_{U_1}(z)\mathrm{id}_{\mathcal{H}_2}$$

holds for all elements $z \in Z$. If the coefficients c_{U_1,f_1,g_1} for $(f_1,g_1) \in \mathcal{H}_1 \times \mathcal{H}_1$ and c_{U_2,f_2,g_2} for $(f_2,g_2) \in \mathcal{H}_2 \times \mathcal{H}_2$ of $(U_1,\mathcal{H}_1)$ and $(U_2,\mathcal{H}_2)$, respectively, belong to the representation space $\mathcal{H} = L^2(G/Z,\bar{\chi}_{U_1};\mu)$ of $\mathrm{Ind}_Z^G(\bar{\chi}_{U_1}.\mathrm{id}_{\mathbb{C}},\mathbb{C})$ - then $\mathcal{R}_G(U_1,U_2) = \{0\}$ implies

$$\langle c_{U_1,f_1,g_1} | c_{U_2,f_2,g_2} \rangle = 0$$

for all pairs $(f_1,g_1) \in \mathcal{H}_1 \times \mathcal{H}_1$ and $(f_2,g_2) \in \mathcal{H}_2 \times \mathcal{H}_2$.

Proof. We may assume that the vectors f_1, g_1, f_2, g_2 are all different from zero. For $u_1 \in \mathcal{H}_1$ consider the continuous $\mathbb{C}$-linear form

$$L_{u_1} : \mathcal{H}_2 \ni u_2 \rightsquigarrow \langle c_{U_2,u_2,g_2} | c_{U_1,u_1,g_1} \rangle \in \mathbb{C}$$

where the bracket $\langle \cdot | \cdot \rangle$ denotes the scalar product of the complex Hilbert space $\mathcal{H}$. There exists a unique element $T(u_1) \in \mathcal{H}_2$ such that

$$L_{u_1}(u_2) = \langle u_2 | T(u_1) \rangle$$

holds for all vectors $u_2 \in \mathcal{H}_2$. The linear mapping T defined via the prescription $T : \mathcal{H}_1 \ni u_1 \rightsquigarrow T(u_1) \in \mathcal{H}_2$ is closed, hence continuous. The identities

$$\begin{aligned}
\langle u_2 | T \circ U_1(y)u_1 \rangle &= \langle c_{U_2,u_2,g_2} | c_{U_1,U_1(y)u_1,g_1} \rangle \\
&= \int_{G/Z} \langle U_2(x)u_2 | g_2 \rangle \cdot \overline{\langle U_1(xy)u_1 | g_1 \rangle} d\mu(\dot{x}) \\
&= \int_{G/Z} \langle U_2(xy^{-1})u_2 | g_2 \rangle \cdot \overline{\langle U_1(x)u_1 | g_1 \rangle} d\mu(\dot{x}) \\
&= \langle c_{U_2,U_2(y^{-1})u_2,g_2} | c_{U_1,u_1,g_1} \rangle \\
&- L_{u_1}(U_2(y^{-1})u_2) \\
&= \langle U_2(y^{-1})u_2 | T(u_1) \rangle \\
&= \langle u_2 | U_2(y) \circ T(u_1) \rangle
\end{aligned}$$

which hold for all elements $y \in G$ and all pairs $(u_1,u_2) \in \mathcal{H}_1 \times \mathcal{H}_2$ show that $T \in \mathcal{R}_G(U_1,U_2)$. Thus $T = 0$ and therefore $L_{f_1}(f_2) = 0$. —

Corollary. Let $(U_1,\mathcal{H}_1)$, $(U_2,\mathcal{H}_2)$ be two linear representations of G that are square integrable modulo the centre C of G, and such that their central characters coincide on the closed subgroup Z of C:

$$\chi_{U_1}|Z = \chi_{U_2}|Z = \chi$$

If $(U_1,\mathcal{H}_1)$ and $(U_2,\mathcal{H}_2)$ are non-isomorphic, then their coefficients c_{U_1,f_1,g_1} and c_{U_2,f_2,g_2}, respectively, are orthogonal in the complex Hilbert space $L^2(G/Z,\bar{\chi};\mu)$ for all vectors f_1,g_1, in $\mathcal{H}_1$ and f_2,g_2 in $\mathcal{H}_2$.

3.6 Let $(U_0,\mathcal{H}_0)$ denote a square integrable linear representation of the unimodular, locally compact, topological group G and $\mathcal{L}_{U_0}$ the *closure* of the vector subspace of $L^2(G)$ that is spanned by the set of coefficients

$$\{c_{U_0,f,g} \mid (f,g) \in \mathcal{H}_0 \times \mathcal{H}_0\}.$$

In view of the identities (cf. Section 3.3 supra)

$$\gamma_G(x)c_{U_0,f,g} = \varepsilon_x * c_{U_0,f,g} = c_{U_0,f,U_0(x)g},$$

$$\delta_G(y)c_{U_0,f,g} = c_{U_0,f,g} * \varepsilon_{y^{-1}} = c_{U_0,U_0(y)f,g}$$

which hold for all pairs $(f,g) \in \mathcal{H}_0 \times \mathcal{H}_0$ and all elements $x \in G$ and $y \in G$, the complex Hilbert space $\mathcal{L}_{U_0}$ is bi-invariant, i.e., $\mathcal{L}_{U_0}$ is stable under the left and right action of G on $L^2(G)$. Consider the continuous, unitary, linear representation $(\gamma_G \times \delta_G, L^2(G))$ of $G \times G$ as defined in Section 1.22. Then $\mathcal{L}_{U_0}$ is stable under the action of $G \times G$ by means of $\gamma_G \times \delta_G$ and for any pair $(f_0,g_0) \in \mathcal{H}_0 \times \mathcal{H}_0$ such that $f_0 \neq 0$, $g_0 \neq 0$ the $(\gamma_G \times \delta_G)$-orbit of $c_{U_0,f_0,g_0} \in \mathcal{L}_{U_0}$ is a total subset of $\mathcal{L}_{U_0}$. Thus the linear subrepresentation $(\gamma_G \times \delta_G)|\mathcal{L}_{U_0}$ of $(\gamma_G \times \delta_G, L^2(G))$ is a topologically cyclic, continuous, unitary, linear representation of $G \times G$.

If $(U_1,\mathcal{H}_1)$, $(U_2,\mathcal{H}_2)$ denote two square integrable linear representations of G which are isomorphic then we have

$$\mathcal{L}_{U_1} = \mathcal{L}_{U_2}.$$

In the other case when $(U_1,\mathcal{H}_1)$, $(U_2,\mathcal{H}_2)$ are non-isomorphic, i.e. $\mathcal{R}_G(U_1,U_2) = \{0\}$, then $\mathcal{L}_{U_1}$ and $\mathcal{L}_{U_2}$ are orthogonal vector subspaces of $L^2(G)$ by virtue of the Corollary of Theorem 3.5. It follows that $\mathcal{L}_{U_0}$ does not depend upon the specific choice of the representative $(U_0,\mathcal{H}_0)$ within its isomorphy class in $\hat{G}$. Moreover, if $(f_j)_{j\in I}$ denotes a *Hilbert basis* of $\mathcal{H}_0$ then the family

$$(\sqrt{\deg(U_0)}\; c_{U_0,f_j,f_k})_{(j,k) \in I \times I}$$

forms a *Hilbert basis* of $\mathcal{L}_{U_0}$.

Let $\mathcal{L}$ denote an arbitrary closed vector subspace of $L^2(G)$ that is *stable* under the linear action of G by means of the left regular representation $(\gamma_G, L^2(G))$ of G. Suppose that the linear subrepresentation $V = \gamma_G|\mathcal{L}$ of G in $\mathcal{L}$ is topologically irreducible. An application of Fubini's theorem yields for any pair $(\phi,\psi) \in \mathcal{L} \times \mathcal{L}$ the identities

$$\begin{aligned}\int_G |c_{V,\phi,\psi}(x)|^2 dx &= \int_G \left(\int_G |\phi(x^{-1}y)\bar{\psi}(y)|^2 dy\right)dx \\ &= \int_G \left(\int_G |\phi(x^{-1}y)|^2 dx\right)|\psi(y)|^2 dy \\ &= \|\phi\|_2^2 \|\psi\|_2^2.\end{aligned}$$

It follows $c_{V,\phi,\psi} \in L^2(G)$ so that $(V,\mathcal{L})$ is a square integrable linear representation of G in the complex Hilbert space $\mathcal{L}$. Let $P_{\mathcal{L}}$ denote the orthoprojector of $L^2(G)$ onto $\mathcal{L}$. If $\phi \in \mathbb{K}(G)$, $\psi \in \mathcal{L}$ and $x \in G$, then Lemma 1.5 gives

$$\begin{aligned}c_{V,P_{\mathcal{L}}\phi,\psi}(x) &= \langle V(x) \circ P_{\mathcal{L}}\phi|\psi\rangle = \langle P_{\mathcal{L}} \circ V(x)\phi|\psi\rangle \\ &= \langle V(x)\phi|\psi\rangle = \int_G \phi(x^{-1}y)\bar{\psi}(y)dy \\ &= \bar{\psi} \star \check{\phi}.\end{aligned}$$

Thus $\bar{c}_{V,P_{\mathcal{L}}\phi,\psi} = \psi \star \phi^*$. If ϕ^* runs through a continuous approximate unit of the complex Banach algebra $L^1(G)$ we obtain $\psi \in \mathcal{L}_{\check{V}}$, where $\check{V} : G \ni x \rightsquigarrow V(x^{-1}) \in \underline{\underline{U}}(\mathcal{L})$. Consequently

$$\mathcal{L} \subseteq \mathcal{L}_{\check{V}}.$$

In other words: The closed vector subspace of $L^2(G)$ that is spanned by the union of the vector subspaces $\mathcal{L}_{U_0}$ where $(U_0,\mathcal{H}_0)$ runs through a family of representatives of the isomorphy classes of square integrable linear representations of G includes all the vector subspaces $\mathcal{L}$ of $L^2(G)$ for which $\gamma_G|\mathcal{L}$ is a topologically irreducible linear subrepresentation of the left regular representation $(\gamma_G, L^2(G))$ of G. Of course, a similar result holds for the right regular representation $(\delta_G, L^2(G))$, too (cf. Section 1.22).

Consider the orthoprojector $P_{U_0} = P_{\mathcal{L}_{U_0}}$ of $L^2(G)$ onto the closed vector subspace $\mathcal{L}_{U_0}$ of $L^2(G)$. If $(f_j)_{j\in I}$ denotes a Hilbert basis of $\mathcal{H}_0$ then we obtain for all elements $\phi \in L^1(G) \cap L^2(G)$

$$\begin{aligned}
\|P_{U_0}(\bar{\phi})\|_2^2 &= \deg(U_0) \sum_{(j,k)\in I\times I} |\langle c_{U_0,f_j,f_k}|\bar{\phi}\rangle|^2 \\
&= \deg(U_0) \sum_{(j,k)\in I\times I} |\langle U_0^1(\phi)f_j|f_k\rangle|^2 \\
&= \deg(U_0) \sum_{j\in I} \|U_0^1(\phi)f_j\|_2^2 .
\end{aligned}$$

From these identities we infer the following result:

3.7 Theorem. Let $(U_0, \mathcal{H}_0)$ denote a square integrable linear representation of the unimodular, locally compact, topological group G and $\phi \in L^1(G) \cap L^2(G)$. Then $U_0^1(\phi) : \mathcal{H}_0 \to \mathcal{H}_0$ is a Hilbert-Schmidt operator with norm

$$\|U_0^1(\phi)\|_2 = \frac{1}{\sqrt{\deg(U_0)}} \|P_{U_0}(\bar{\phi})\|_2 .$$

A topological group G is called to be a group which admits *arbitrarily small invariant neighbourhoods* if any given neighbourhood of its neutral element 1_G contains a neighbourhood N of 1_G in G such that N is stable under the action of G by the inner automorphisms $\mathrm{Int}_G(x)$ of G induced by the elements $x \in G$, i.e., the inclusion

$$\mathrm{Int}_G(x)N \subseteqq N$$

holds for every $x \in G$. A covering argument shows that every *compact* topological group G admits arbitrarily small invariant neighbourhoods.

Corollary. Every square integrable linear representation $(U_0, \mathcal{H}_0)$ of a unimodular, locally compact, topological group G which admits arbitrarily small invariant neighbourhoods is finite dimensional, i.e.,

$$\dim_{\mathbb{C}} \mathcal{H}_0 < +\infty .$$

Proof. Let N be relatively compact neighbourhood of the neutral element 1_G in G which is invariant under the inner automorphisms $\mathrm{Int}_G(x)$ of G induced by the elements $x \in G$ and let $\phi_N : G \to \mathbb{C}$ denote its indicator function. Then $\phi_N \in L^1(G) \cap L^2(G)$ belongs to the centre of the involutory complex Banach algebra $L^1(G)$ and Theorem 1.18 implies

$$U_o^1(\phi_N) = c_N . \mathrm{id}_{\mathcal{H}_o}$$

where c_N denotes a complex number. Let $f_o \in \mathcal{H}_o$ be any vector of norm $||f_o|| = 1$. Then we have

$$c_N = \langle U_o^1(\phi_N) f_o | f_o \rangle = \int_N c_{U_o, f_o, f_o}(x) dx.$$

By choosing N to be sufficiently small we see that $c_N \neq 0$. Since $U_o^1(\phi_N)$ is a Hilbert-Schmidt operator of $\mathcal{H}_o$ into itself we conclude $\dim_{\mathbb{C}} \mathcal{H}_o < +\infty$. —

3.8 The results of Section 3.6 can be reformulated in a slightly different but nevertheless useful way. In order to do this appropriately, we shall summarize some generalities on tensor products of linear group representations. Let $(U_1, \mathcal{H}_1)$ and $(U_2, \mathcal{H}_2)$ denote two continuous, unitary, linear representations of the unimodular, locally compact, topological groups G_1 and G_2, respectively. Form the *tensor product* $\mathcal{H}_1 \otimes \mathcal{H}_2$ and let the complex vector space $\mathcal{H}_1 \otimes \mathcal{H}_2$ be equipped with its natural scalar product $\langle \cdot | \cdot \rangle_2$. Observe that $\langle \cdot | \cdot \rangle_2$ is well defined via the prescription

$$\langle f_1 \otimes f_2 | g_1 \otimes g_2 \rangle_2 = \langle f_1 | g_1 \rangle . \langle f_2 | g_2 \rangle .$$

The *completion* $\mathcal{H}_1 \hat{\otimes}_2 \mathcal{H}_2$ of the complex prehilbert space $\mathcal{H}_1 \otimes \mathcal{H}_2$ under the norm $|| \cdot ||_2$ associated with the extension of its natural scalar product $\langle \cdot | \cdot \rangle_2$ forms a complex Hilbert space. There is a natural linear mapping

$$\underline{\underline{U}}(\mathcal{H}_1) \times \underline{\underline{U}}(\mathcal{H}_2) \ni (S,T) \rightsquigarrow S \hat{\otimes}_2 T \in \underline{\underline{U}}(\mathcal{H}_1 \hat{\otimes}_2 \mathcal{H}_2).$$

In particular, the assignment

$$G_1 \times G_2 \ni (x_1, x_2) \rightsquigarrow U_1(x_1) \hat{\otimes}_2 U_2(x_2)$$

defines a continuous, unitary, linear representation $(U_1 \hat{\otimes}_2 U_2, \mathcal{H}_1 \hat{\otimes}_2 \mathcal{H}_2)$ of

$G_1 \times G_2$ which is called to be the *tensor product* of $(U_1,\mathcal{H}_1)$ and $(U_2,\mathcal{H}_2)$. Fix a pair of vectors $(f_1,f_2) \in \mathcal{H}_1 \times \mathcal{H}_2$ and consider the continuous $\mathbb{C}$-linear mapping T_{f_1,f_2} defined via the prescription

$$T_{f_1,f_2} : \mathcal{H}_1 \ni f \rightsquigarrow \langle f|f_1\rangle f_2 \in \mathcal{H}_2.$$

Then T_{f_1,f_2} forms a *dyad* and the $\mathbb{C}$-sesquilinear mapping

$$(f_1,f_2) \rightsquigarrow T_{f_1,f_2}$$

give rise to an isomorphism Φ of $\mathcal{H}_1 \otimes \mathcal{H}_2$ onto the complex vector space $\mathcal{L}_f(\mathcal{H}_1,\mathcal{H}_2)$ of all $\mathbb{C}$-linear mappings of $\mathcal{H}_1$ onto $\mathcal{H}_2$ with *finite rank*. Transport by means of Φ the prehilbert structure of the complex vector space $\mathcal{H}_1 \otimes \mathcal{H}_2$ given by $\langle\cdot|\cdot\rangle_2$ to $\mathcal{L}_f(\mathcal{H}_1,\mathcal{H}_2)$. Observe that the completion of the complex prehilbert space $\mathcal{L}_f(\mathcal{H}_1,\mathcal{H}_2)$ is the complex Hilbert space $\mathcal{L}_2(\mathcal{H}_1,\mathcal{H}_2)$ of all *Hilbert-Schmidt operators* of $\mathcal{H}_1$ into $\mathcal{H}_2$. The complex Hilbert space $\mathcal{L}_2(\mathcal{H}_1,\mathcal{H}_2)$ contains the vector subspace $\mathcal{L}_1(\mathcal{H}_1,\mathcal{H}_2)$ of all continuous $\mathbb{C}$-linear mappings of $\mathcal{H}_1$ into $\mathcal{H}_2$ of *trace class*. Thus we have the filtration

$$\mathcal{L}_f(\mathcal{H}_1,\mathcal{H}_2) \hookrightarrow \mathcal{L}_1(\mathcal{H}_1,\mathcal{H}_2) \hookrightarrow \mathcal{L}_2(\mathcal{H}_1,\mathcal{H}_2)$$

of complex vector spaces of *compact* $\mathbb{C}$-linear mappings of $\mathcal{H}_1$ into $\mathcal{H}_2$. The isomorphism $\Phi:\mathcal{H}_1 \otimes \mathcal{H}_2 \to \mathcal{L}_f(\mathcal{H}_1,\mathcal{H}_2)$ admits a unique extension $\hat{\Phi}$ to a unitary isomorphism of the complex Hilbert space $\mathcal{H}_1 \hat{\otimes}_2 \mathcal{H}_2$ onto the complex Hilbert space $\mathcal{L}_2(\mathcal{H}_1,\mathcal{H}_2)$. Moreover, $\hat{\Phi}$ is a unitary *isomorphism* of $(U_1 \hat{\otimes}_2 U_2, \mathcal{H}_1 \hat{\otimes}_2 \mathcal{H}_2)$ onto the continuous, unitary, linear representation

$$V:(G_1\times G_2 \ni (x_1,x_2) \rightsquigarrow (\mathcal{L}_2(\mathcal{H}_1,\mathcal{H}_2) \ni T \rightsquigarrow U_2(x_2) \circ T \circ U_1(x_1^{-1}) \in \mathcal{L}_2(\mathcal{H}_1,\mathcal{H}_2))$$

of $G_1 \times G_2$ in the complex Hilbert space $\mathcal{L}_2(\mathcal{H}_1,\mathcal{H}_2)$ of all Hilbert-Schmidt operators of $\mathcal{H}_1$ into $\mathcal{H}_2$. In the following we shall identify the continuous, unitary, linear representations $(\check{U}_1 \hat{\otimes}_2 U_2, \mathcal{H}_1 \hat{\otimes}_2 \mathcal{H}_2)$ and $(V,\mathcal{L}_2(\mathcal{H}_1,\mathcal{H}_2))$ of $G_1 \times G_2$ by means of the unitary isomorphism $\hat{\Phi}$. In the case when $(U_1,\mathcal{H}_1)$ and $(U_2,\mathcal{H}_2)$ are topologically irreducible, continuous, unitary, linear representations of G_1 and G_2 in $\mathcal{H}_1$ and $\mathcal{H}_2$, respectively, then $(\check{U}_1 \hat{\otimes}_2 U_2, \mathcal{H}_1 \hat{\otimes}_2 \mathcal{H}_2)$ and $(V, \mathcal{L}_2(\mathcal{H}_1,\mathcal{H}_2))$ are topologically irreducible, continuous, unitary, linear representations of $G_1 \times G_2$. The restrictions of $(\check{U}_1 \hat{\otimes}_2 U_2,$

$\mathcal{H}_1 \hat{\otimes}_2 \mathcal{H}_2$) onto the closed subgroups $G_1 \times \{1_{G_2}\}$ and $\{1_{G_1}\} \times G_2$ of $G_1 \times G_2$ are multiplies of $(\check{U}_1,\mathcal{H}_1)$ and $(U_2,\mathcal{H}_2)$, respectively.

3.9 Let $(U_0,\mathcal{H}_0)$ denote a topologically irreducible, continuous, unitary, linear representation of the unimodular, locally compact, topological group G in the complex Hilbert space $\mathcal{H}_0$. Put $G_1 = G_2 = G$, $U_1 = U_2 = U_0$, $\mathcal{H}_1 = \mathcal{H}_2 = \mathcal{H}_0$, and $\mathcal{L}_2(\mathcal{H}_0,\mathcal{H}_0) = \mathcal{L}_2(\mathcal{H}_0)$. Observe that $\mathcal{L}_2(\mathcal{H}_0)$ forms a *complete Hilbert algebra* over $\mathbb{C}$. The continuous, unitary, linear representation $(id_{\mathcal{H}_0} \hat{\otimes}_2 U_0, \mathcal{L}_2(\mathcal{H}_0))$ of G in the complex Hilbert space of Hilbert-Schmidt operators of $\mathcal{H}_0$ is a multiple of $(U_0,\mathcal{H}_0)$. Similarly, the continuous unitary linear representation $(\check{U}_0 \hat{\otimes}_2 id_{\mathcal{H}_0}, \mathcal{L}_2(\mathcal{H}_0))$ of G is a multiple of the contragredient representation $(\check{U}_0,\mathcal{H}_0)$ of G.

Suppose that $(U_0,\mathcal{H}_0)$ denotes a square integrable, linear, representation of G in the complex Hilbert space $\mathcal{H}_0$. The identity

$$(\gamma_G \times \delta_G)(x,y)\, c_{U_0,f,g} = c_{U_0,U_0(y)f,U_0(x)g}$$

which holds for all pairs $(f,g) \in \mathcal{H}_0 \times \mathcal{H}_0$ and all pairs $(x,y) \in G \times G$ shows that the $\mathbb{C}$-linear mapping defined via the prescription

$$\mathcal{H}_0 \otimes \mathcal{H}_0 \ni f \otimes g \rightsquigarrow c_{U_0,f,g} \in \mathcal{L}_{U_0}$$

gives rise to an isomorphism of $(V,\mathcal{L}_2(\mathcal{H}_0))$ and hence of $(\check{U}_0 \hat{\otimes}_2 U_0, \mathcal{H}_0 \hat{\otimes}_2 \mathcal{H}_0)$ onto the topologically irreducible, continuous, unitary, linear representation $(\gamma_G \times \delta_G, \mathcal{L}_{U_0})$ of $G \times G$ in the complex Hilbert space $\mathcal{L}_{U_0}$.

Conversely, suppose that $(U_0,\mathcal{H}_0)$ denotes a topologically irreducible, continuous, unitary, linear representation of G in the complex Hilbert space $\mathcal{H}_0$ such that the $\mathbb{C}$-linear mapping

$$K(G) \ni \phi \rightsquigarrow U_0^1(\phi) \in \mathrm{End}(\mathcal{H}_0)$$

extends to a unitary isomorphism of $L^2(G)$ onto $\mathcal{L}_2(\mathcal{H}_0)$, then $(U_0,\mathcal{H}_0)$ is a square integrable, linear representation of G.

3.10 Theorem. Let G be a unimodular, locally compact, topological group and $(U_1,\mathcal{H}_1)$, $(U_2,\mathcal{H}_2)$ two continuous, unitary, linear representations of G such that for all functions $\phi \in \mathbb{K}(G)$ the continuous $\mathbb{C}$-linear mappings

$$U_1^1(\phi) : \mathcal{H}_1 \to \mathcal{H}_1,$$

$$U_2^1(\phi) : \mathcal{H}_2 \to \mathcal{H}_2,$$

belong to the complex Banach spaces $\mathcal{L}_2(\mathcal{H}_1)$ and $\mathcal{L}_2(\mathcal{H}_2)$, respectively, and satisfy the equality

$$||U_2^1(\phi)||_2^2 = \rho ||U_1^1(\phi)||_2^2$$

with a fixed real number $\rho > 0$ for all functions $\phi \in \mathbb{K}(G)$. If $(U_1, \mathcal{H}_1)$ is topologically irreducible, then

$$\rho = n \in \mathbb{N}$$

and $(U_1, \mathcal{H}_1)$ is contained in $(U_2, \mathcal{H}_2)$ with multiplicity n, i.e., we have

$$U_2 = n.U_1.$$

Proof. Define the $\mathbb{C}$-linear mapping

$$M : U_1^1(\mathbb{K}(G)) \ni U_1^1(\phi) \rightsquigarrow U_2^1(\phi) \in \mathcal{L}_2(\mathcal{H}_2)$$

and extend it uniquely from the everywhere dense vector subspace $U_1^1(\mathbb{K}(G))$ of $\mathcal{L}_2(\mathcal{H}_1)$ to a representation

$$\mathcal{L}_2(\mathcal{H}_1) \ni T \rightsquigarrow M(T) \in \mathcal{L}_2(\mathcal{H}_2)$$

of the involutory complex Banach algebra $\mathcal{L}_2(\mathcal{H}_1)$ in the complex Hilbert space $\mathcal{H}_2$. In view of the identity

$$||M(T)||_2^2 = \rho ||T||_2^2$$

which holds for all operators $T \in \mathcal{L}_2(\mathcal{H}_1)$, the representation M of $\mathcal{L}_2(\mathcal{H}_1)$ in $\mathcal{H}_2$ is faithful. The image $M(\mathcal{L}_2(\mathcal{H}_1))$ forms a closed selfadjoint subalgebra of the complete Hilbert algebra $\mathcal{L}_2(\mathcal{H}_2)$ over $\mathbb{C}$. The minimal left ideals of the complete Hilbert algebra $M(\mathcal{L}_2(\mathcal{H}_1))$ over $\mathbb{C}$, i.e., the minimal closed vector subspaces $\neq \{0\}$ of $M(\mathcal{L}_2(\mathcal{H}_1))$ which are stable under the action of $id_{\mathcal{H}_2} \hat{\otimes}_2 U_2$ have the same finite dimension $n = \rho \in \mathbb{N}$ over $\mathbb{C}$. Take such a minimal left ideal $\mathcal{I}$ of $M(\mathcal{L}_2(\mathcal{H}_1))$ and form the tensor product $\mathcal{H}_1 \hat{\otimes}_2 \mathcal{I}$.

Since the continuous, unitary, linear representation $(n.U_1, H_1 \hat{\otimes}_2 1)$ of G is unitarily isomorphic to (U_2, H_2) by means of M, the result follows.—

3.11 Let (U,H) denote a continuous, unitary, linear representation of the unimodular, locally compact, topological group G in the complex Hilbert space H. Denote by $\mathcal{L}_1(H)$ the complex vector space of all continuous $\mathbb{C}$-linear mappings of H into itself which are of *trace class*. Then we have the inclusion maps

$$\mathcal{L}_f(H) \hookrightarrow \mathcal{L}_1(H) \hookrightarrow \mathcal{L}_2(H)$$

and $\mathcal{L}_1(H)$ forms the completion of the vector space $\mathcal{L}_f(H) = \mathcal{L}_f(H,H)$ under the *trace norm*

$$\mathcal{L}_f(H) \ni T \rightsquigarrow \|T\|_1 = tr(|T|).$$

(cf. Section 1.3). Suppose that for all functions $\phi \in K(G)$ the continuous $\mathbb{C}$-linear mapping

$$U^1(\phi) : H \to H$$

belongs to the Banach space $\mathcal{L}_1(H)$. Then (U,H) is called to be a *unitary linear representation of trace class* of G in the complex Hilbert space H. Theorem 3.10 supra will be applied in Section 6 to topologically irreducible, unitary, linear representations of trace class.

References

Gaal, S.A. : Linear analysis and representation theory. Die Grundlehren der mathematischen Wissenschaften, Band 198. Springer, Berlin, Heidelberg, New York 1973.

Godement, R. : Sur les relations d'orthogonalité de V. Bargmann. I. Résultats préliminaires. C.R. Acad. Sci. Paris 225 (1947), 521-523.

Godement, R. : Sur les relations d'orthogonalité de V. Bargmann. II. Démonstration générale. C.R. Acad. Sci. Paris 225 (1947), 657-659.

Moore, C.C., Wolf, J.A. : Square integrable representations of nilpotent groups. Trans. Amer. Math. Soc. 185 (1973), 445-462.

Rieffel, M.A. : Square integrable representations of Hilbert algebras.
J. Funct. Anal. 3 (1969), 265-300.

Shucker, D.S. : Square integrable representations of unimodular groups.
Proc. Amer. Math. Soc. 89 (1983), 169-172.

4 Basic facts on real nilpotent Lie groups

4.1 Let G denote a group that is written multiplicatively and let $1_G \in G$ be its neutral element. For any two subsets A, B of G we denote by [A,B] the subgroup of G that is generated by the set of commutators

$$\{[x,y] \mid (x,y) \in A \times B\}$$

where, as usual, $[x,y] = (xy).(yx)^{-1} = xyx^{-1}y^{-1}$. Note that if A and B are normal subgroups of G, then [A,B] is a normal subgroup of G. In particular, the *derived* group of G (cf. Section 4.5 infra),

$$\mathfrak{D}^1 G = [G,G],$$

generated by the commutators of G, forms a *normal* subgroup of G.

Let $p \in \mathbb{N}$ be a natural number. Define the *descending central series* $(\mathfrak{C}^p G)_{p \geqq 0}$ of G recursively via the prescriptions

$$\begin{cases} \mathfrak{C}^0 G = G, \\ \mathfrak{C}^{p+1} G = [G, \mathfrak{C}^p G]. \end{cases}$$

Then we get the following descending filtration of *normal subgroups* of G:

$$G \hookleftarrow \mathfrak{C}^1 G \hookleftarrow \mathfrak{C}^2 G \hookleftarrow \ldots \hookleftarrow \mathfrak{C}^p G \hookleftarrow \ldots \hookleftarrow \{1_G\}$$

The group G is called to be *nilpotent* if the sequence $(\mathfrak{C}^p G)_{p \geqq 0}$ is finite and terminates at $\{1_G\}$, i.e., if there exists a sufficiently large natural number p_0 such that

$$\mathfrak{C}^{p_0} G = \{1_G\}$$

holds. If $p_0 \geqq 1$ and $\mathfrak{C}^{p_0 - 1} G \neq \{1_G\}$ then the number p_0 is called to be the *length* of the nilpotent group G and G is called to be a p_0-*step nilpotent* group.

Remark. The one-step nilpotent groups G are formed by the abelian groups

$G \neq \{1_G\}$. If in the category of nilpotent groups the level of hierarchy is measured by their lengths, then the Heisenberg groups belong to the second level. Indeed, they belong to the categorie of 2-step nilpotent real Lie groups (cf. Section 5.4 infra).

4.2 The *ascending central series* $(\mathfrak{C}_p G)_{p \geq 0}$ of the group G is defined recursively according to the rules

$$\begin{cases} \mathfrak{C}_0 G = \{1_G\}, \\ \mathfrak{C}_{p+1} G = \text{preimage of the centre of } G/\mathfrak{C}_p(G) \text{ under the canonical} \\ \qquad\qquad \text{epimorphism } G \to G/\mathfrak{C}_p(G). \end{cases}$$

Thus $\mathfrak{C}_1 G$ is the *centre* C of G, we have

$$\mathfrak{C}_{p+1} G = \{x \in G \mid [x,y] \in \mathfrak{C}_p(G) \textit{ for all } y \in G\},$$

and the following ascending filtration of *normal subgroups* of G arises:

$$\{1_G\} \hookrightarrow \mathfrak{C}_1 G \hookrightarrow \mathfrak{C}_2 G \hookrightarrow \ldots \hookrightarrow \mathfrak{C}_p G \hookrightarrow \ldots \hookrightarrow G$$

4.3 Lemma. Let $(G_p)_{0 \leq p \leq p_0}$ be a sequence of subgroups of the group G such that $G_0 = G$, and

$$G_0 \hookleftarrow G_1 \hookleftarrow G_2 \hookleftarrow \ldots \hookleftarrow G_p \hookleftarrow \ldots \hookleftarrow G_{p_0},$$

and

$$[G,G_p] \subseteqq G_{p+1} \quad (0 \leq p \leq p_0 - 1).$$

Then $\mathfrak{C}^p G \subseteqq G_p$ for $0 \leq p \leq p_0$. If $G_{p_0} = \{1_G\}$ - then $G_p \subseteqq \mathfrak{C}_{p_0 - p} G$ for $0 \leq p \leq p_0$ and $\mathfrak{C}_{p_0} G = G$.

Proof. In the case $p = 0$ we have $\mathfrak{C}^0 G = G = G_0$. For $p = 1$ we obtain

$$\mathfrak{C}^1 G = [G,G] = [G,G_0] \subseteqq G_1$$

by the hypothesis. Assume $\mathfrak{C}^p G \subseteqq G_p$ for any natural number p such that $0 \leq p \leq p_0 - 1$. Then we obtain

$$\mathfrak{C}^{p+1} G = [G,\mathfrak{C}^p G] \subseteqq [G,G_p] \subseteqq G_{p+1}.$$

If $G_{p_0} = \{1_G\}$ then $[G,G_{p_0-1}] = \{1_G\}$ and therefore $G_{p_0-1} \subseteqq C = \mathfrak{C}_1 G$. If we have $G_p \subseteqq \mathfrak{C}_{p_0-p} G$ for a natural number p satisfying $0 \leqq p \leqq p_0$, then $[G,G_{p-1}] \subseteqq G_p$ implies $G_{p-1} \subseteqq \mathfrak{C}_{p_0-p+1} G$. Finally $G_0 \subseteqq \mathfrak{C}_{p_0} G \subseteqq G$ implies $\mathfrak{C}_{p_0} G = G$. —

4.4 <u>Theorem</u>. Let G be a group with neutral element 1_G and let $p_0 \geqq 1$ be a sufficiently large natural number. The following conditions are pairwise equivalent:

(i) $\mathfrak{C}^{p_0} G = \{1_G\}$,

(ii) There exists a sequence $(G_p)_{0 \leqq p \leqq p_0}$ of subgroups of G such that $G_0 = G$, $G_{p_0} = \{1_G\}$, and

$$G_0 \hookleftarrow G_1 \hookleftarrow G_2 \hookleftarrow \dots \hookleftarrow G_p \hookleftarrow \dots \hookleftarrow G_{p_0},$$

and

$$[G,G_p] \subseteqq G_{p+1} \quad (0 \leqq p \leqq p_0-1),$$

(iii) $\mathfrak{C}_{p_0} G = G$.

Thus G is a nilpotent group if it satisfies one, hence all of the conditions (i), (ii), (iii) for a sufficiently large natural number p_0.

<u>Proof</u>. If we set $G_p = \mathfrak{C}^p G$ for $0 \leqq p \leqq p_0$ then (i) implies the condition (ii). Moreover (ii) implies (iii) by Lemma 4.3. Suppose that G satisfies the condition (iii) - then

$$\mathfrak{C}^1 G = [G,G] = [\mathfrak{C}_{p_0} G, \mathfrak{C}_{p_0} G] \subseteqq \mathfrak{C}_{p_0-1} G.$$

Now suppose $\mathfrak{C}^p G \subseteqq \mathfrak{C}_{p_0-p} G$ for a natural number p such that $0 \leqq p \leqq p_0-1$. Then we have

$$\mathfrak{C}^{p+1} G = [G,\mathfrak{C}^p G] \subseteqq [G,\mathfrak{C}_{p_0-p} G].$$

Consequently

$$\mathfrak{C}^{p+1} G \subseteqq \mathfrak{C}_{p_0-p-1} G,$$

and therefore $\mathfrak{C}^{p_0}G = \{1_G\}$. Thus G satisfies the condition (i). —

Corollary 1. Let G denote a nilpotent group.

(i) Each subgroup H of G is nilpotent.

(ii) For each normal subgroup H of G the quotient group G/H is nilpotent.

Proof. Let $p_0 \geqq 1$ denote the length of G. In the case (i) set

$$G_p = H \cap \mathfrak{C}^p G \quad (0 \leqq p \leqq p_0).$$

In the case (ii) let $\pi: G \to G/H$ be the canonical epimorphism and let

$$G_p = \pi(\mathfrak{C}^p G) \qquad (0 \leqq p \leqq p_0)$$

be the image of $\mathfrak{C}^p G$ under π. An application of Theorem 4.4 yields the result in both cases. —

Corollary 2. Let $G \neq \{1_G\}$ be a nilpotent group with centre C - then $C \neq \{1_G\}$.

Theorem 4.4 supra represents a convenient criterion to establish the nilpotency of certain groups. In Lemma 5.3 infra it will serve to establish that the automorphism group $T_a(\mathcal{F},\mathbb{R})$, $a \neq 0$, of any flag $\mathcal{F}$ in a finite dimensional real vector space E forms a nilpotent real Lie group (cf. 5.1 infra).

4.5 The *derived series* $(\mathfrak{D}^p G)_{p \geqq 0}$ of the group G is defined recursively via the prescriptions

$$\begin{cases} \mathfrak{D}^0 G = G, \\ \mathfrak{D}^{p+1} G = [\mathfrak{D}^p G, \mathfrak{D}^p G] \end{cases}$$

and yields the descending filtration of *normal subgroups* of G

$$G \hookleftarrow \mathfrak{D}^1 G \hookleftarrow \mathfrak{D}^2 G \hookleftarrow \dots \hookleftarrow \mathfrak{D}^p G \hookleftarrow \dots \hookleftarrow \{1_G\}.$$

Note that $\mathfrak{D}^1 G = \mathfrak{C}^1 G = [G,G]$ is the derived group of G (Section 4.1). The group G is called to be *solvable* if there exists a sufficiently large natural number $p_0 \geqq 0$ such that

$$\mathfrak{D}^{p_0} G = \{1_G\}$$

holds. If $p_0 \geqq 1$ and $\mathfrak{D}^{p_0 - 1} G \neq \{1_G\}$ then the number p_0 is called to be the *length* of the solvable group G. Observe that nilpotency of the group G involves a descending filtration using commutators of the terms of the filtration with G, whereas solvability of G involves a descending filtration using commutators of the terms of the filtration with itself. Obviously we have $\mathfrak{C}^p G \supseteqq \mathfrak{D}^p G$ for all natural numbers p. Thus *every nilpotent group* G *is solvable*, the converse being false. Since the Heisenberg groups are nilpotent real Lie groups (cf. Section 5.4 infra) there is no need for us to study in some detail the larger category of solvable Lie groups. The following remarkable fact concerning finite dimensional, linear representations of solvable groups, however, can be easily established.

4.6 Theorem. Let G be a connected, solvable, locally compact, topological group and $(U, \mathfrak{H})$ a finite dimensional, irreducible, continuous, unitary, linear representation of G - then $\dim_{\mathbb{C}} \mathfrak{H} = 1$.

Proof. Let $p_0 \geqq 1$ be the length of the solvable group G and define the connected, closed, normal subgroups $(G_p)_{0 \leqq p \leqq p_0}$ of G via the prescription

$$G_p = \overline{\mathfrak{D}^p G} \qquad (0 \leqq p \leqq p_0).$$

Then $G_0 = G$, $G_{p_0} = \{1_G\}$, and G_p is a closed normal subgroup of G_{p-1} for $1 \leqq p \leqq p_0$. The theorem will be established by induction on p_0. In the case $p_0 = 1$, i.e., when G is abelian the result follows from Corollary 1 of Theorem 1.6. Let now $p_0 > 1$ and assume that the theorem already has been established for connected, solvable, locally compact, topological groups of length $\leqq p_0 - 1$. For any finite dimensional, irreducible, continuous, unitary, linear representation $(U, \mathfrak{H})$ of G decompose the linear representation which is subduced by $(U, \mathfrak{H})$ on the closure G_1 of the derived group of G into a direct sum (Section 1.4) of irreducible, continuous, unitary, linear representations of G_1. By the induction hypothesis, the irreducible components have dimension 1 over $\mathbb{C}$. Let $\hat{G}_1^1$ denote the set of continuous unitary characters of the locally compact topological group G_1. Let $\hat{G}_1^1$ be equipped with the topology of uniform convergence on the compact subsets of G_1. For each character $\chi \in \hat{G}_1^1$ define the set

$$\mathfrak{H}_\chi = \{f \in \mathfrak{H} \mid U(x_1)f = \chi(x_1)f \text{ for all } x_1 \in G_1\}.$$

Moreover, let

$$M_1 = \{\chi \in \hat{G}_1^1 | \mathcal{H}_\chi \neq \{0\}\}.$$

It is easy to see that the family of vector subspaces $(\mathcal{H}_\chi)_{\chi \in M_1}$ of $\mathcal{H}$ are linearly independent over $\mathbb{C}$. It follows that M_1 is a finite subset of $\hat{G}_1^1$. Let $\chi \in M_1$ be arbitrary. Then we obtain for all vectors $f \in \mathcal{H}_\chi$ the identity

$$U(x_1) \circ U(x)f = U(x) \circ U(x^{-1}x_1x)f = \chi(x^{-1}x_1x)U(x)f$$

where $x_1 \in G_1$ and $x \in G$. Define the character of G_1

$$\chi_x = \chi \circ \mathrm{Int}_G(x^{-1})$$

for all $x \in G$. It follows

$$U(x)(\mathcal{H}_\chi) \subseteq \mathcal{H}_{\chi_x}$$

and

$$\chi_x \in M_1$$

for all elements $x \in G$. Since the action $G \ni x \quad \chi_x \in M_1$ of the connected topological group G on the discrete topological subspace M_1 of $\hat{G}_1^1$ is continuous, it must be trivial. Thus the identity

$$\chi_x = \chi_{1_G} = \chi \qquad (x \in G)$$

holds for the previously fixed character $\chi \in M_1$ of G_1. The irreducibility of $(U,\mathcal{H})$ implies the identities

$$\mathcal{H}_\chi = \mathcal{H}$$

and

$$U|G_1 = \chi \mathrm{id}_{\mathcal{H}}.$$

Let us now take an element $x \in G$ and an eigenvector $f \neq 0$ of $U(x) \in \underline{\underline{U}}(\mathcal{H})$ in $\mathcal{H}$, say

$$U(x)f = \zeta f,$$

where $\zeta \in \mathbb{T}$. From the relation

$$U(x) \circ U(y) = \chi([x^{-1},y^{-1}])\, U(y) \circ U(x)$$

which holds for all elements $y \in G$, it follows that $U(y)f$ is also an eigenvector of $U(x)$ corresponding to the eigenvalue $\zeta\chi([x^{-1},y^{-1}]) \in \mathbb{T}$. But the spectrum of $U(x) \in \underline{\underline{U}}(\mathcal{H})$ is a finite subset of $\mathbb{T}$ and the action $G \ni y \rightsquigarrow \zeta\chi([x^{-1},y^{-1}])$ of G on the discrete spectrum of $U(x)$ is continuous. As before, the connectedness of G implies that $\zeta\chi([x^{-1},y^{-1}])$ must be independent of y and therefore equal to $\zeta\chi([x^{-1},1_G]) = \zeta$. This proves that the eigenspace of $U(x)$ corresponding to the eigenvalue $\zeta \in \mathbb{T}$ is stable under G (and $\neq\{0\}$) hence equal to the whole representation space $\mathcal{H}$ by virtue of the irreducibility of $(U,\mathcal{H})$. We have

$$U(x) = \zeta\, id_{\mathcal{H}}$$

for all $x \in G$. Thus G acts by homotheties on $\mathcal{H}$ and by the irreducibility of $(U,\mathcal{H})$ we conclude that $\dim_{\mathbb{C}}\mathcal{H} = 1$. —

In view of the preceding result the unitary dual $\hat{G}$ of a connected, solvable, locally compact, topological group G consists of two types of isomorphy classes, namely of

(I) continuous, unitary characters of G,

and

(II) isomorphy classes of infinite dimensional, topologically irreducible, continuous, unitary, linear representations of G.

It will be the purpose of Section 7 infra to make this rough classification complete for the case of the real Heisenberg nilpotent Lie group.

Corollary. A compact, connected, solvable, topological group G is abelian.

Proof. In view of the Corollary of Theorem 3.7 every topologically irreducible, continuous, unitary linear representation of G is one-dimensional. It follows that G is a commutative group. —

Thus in the above classification of the unitary dual $\hat{G}$ of a connected, solvable, locally compact, topological group G the continuous, unitary, linear representations of type (I) exhaust $\hat{G}$ for all compact groups G.

4.7 Let $\mathfrak{g}$ denote a *Lie algebra* over, say a commutative ring k with unit element. Recall that $\mathfrak{g}$ is a unitary k-module equipped with a k-bilinear mapping, i.e., a k-morphism $\mathfrak{g} \otimes_k \mathfrak{g} \to \mathfrak{g}$ denoted by

$$\mathfrak{g} \times \mathfrak{g} \ni (X,Y) \rightsquigarrow [X,Y] \in \mathfrak{g}$$

such that $[X,X] = 0$ holds for all elements $X \in \mathfrak{g}$, and such that the bracket $[\cdot,\cdot]$ satisfies the *Jacobi identity*

$$[X,[Y,Z]] + [Y,[Z,X]] + [Z,[X,Y]] = 0$$

for all elements X,Y,Z in $\mathfrak{g}$.

Thus the k-morphism $\mathfrak{g} \otimes_k \mathfrak{g} \to \mathfrak{g}$ admits the factorization

$$\mathfrak{g} \otimes_k \mathfrak{g} \to \Lambda^2 \mathfrak{g} \to \mathfrak{g}.$$

For all $X \in \mathfrak{g}$ we denote by $ad_{\mathfrak{g}}X$ the *adjoint* k-linear mapping $\mathfrak{g} \in Y \rightsquigarrow [X,Y] \in \mathfrak{g}$ of X. Then a k-submodule $\mathfrak{h}$ of $\mathfrak{g}$ is a *Lie sub-algebra* of $\mathfrak{g}$, resp., an *ideal* in $\mathfrak{g}$ if and only if $\mathfrak{h}$ is stable under all the adjoint linear mappings $ad_{\mathfrak{g}}X$ of $X \in \mathfrak{h}$, resp., $X \in \mathfrak{g}$. It should be observed that each ideal in $\mathfrak{g}$ is automatically a *two-sided* ideal in $\mathfrak{g}$ because of the identity

$$[X,Y] = - [Y,X]$$

which holds for all pairs $(X,Y) \in \mathfrak{g} \times \mathfrak{g}$ and which is a consequence of the fact that the k-bilinear mapping $[\cdot,\cdot]$ of $\mathfrak{g} \times \mathfrak{g}$ in $\mathfrak{g}$ is *alternating*. If $\mathfrak{a},\mathfrak{h}$ are submodules of $\mathfrak{g}$, then $[\mathfrak{a},\mathfrak{h}]$ denotes the image of $\mathfrak{a} \otimes_k \mathfrak{h}$ under the mapping $(X \otimes Y) \rightsquigarrow [X,Y]$, i.e., the submodule of $\mathfrak{g}$ generated by the set

$$\{[X,Y] \mid (X,Y) \in \mathfrak{a} \times \mathfrak{h}\}.$$

In particular, if $\mathfrak{a}$ and $\mathfrak{h}$ are ideals of $\mathfrak{g}$ then $[\mathfrak{a},\mathfrak{h}]$ also is an ideal of $\mathfrak{g}$.

Two elements X,Y of the Lie algebra $\mathfrak{g}$ over k are said to be *commuting*, if $[X,Y] = 0$. The *centre* $\mathfrak{c}$ of $\mathfrak{g}$ consists of all elements $X \in \mathfrak{g}$ that are commuting with all elements $Y \in \mathfrak{g}$. The Jacobi identity shows at once that $\mathfrak{c}$ is actually an ideal in $\mathfrak{g}$. The *centralizer* of a subset L of $\mathfrak{g}$ is $\{X \in \mathfrak{g} \mid [X,L] = 0\}$. Again by the Jacobi identity, the centralizer of L is a Lie subalgebra of $\mathfrak{g}$.

Define the *descending central series* $(\mathfrak{C}^p\mathfrak{g})_{p \geq 0}$ of $\mathfrak{g}$ recursively via the prescriptions

$$\begin{cases} \mathfrak{C}^0\mathfrak{g} = \mathfrak{g}, \\ \mathfrak{C}^{p+1}\mathfrak{g} = [\mathfrak{g},\mathfrak{C}^p\mathfrak{g}]. \end{cases}$$

Then we get the following descending filtration of *ideals* of $\mathfrak{g}$:

$$\mathfrak{g} \hookleftarrow \mathfrak{C}^1\mathfrak{g} \hookleftarrow \mathfrak{C}^2\mathfrak{g} \hookleftarrow \ldots \hookleftarrow \mathfrak{C}^p\mathfrak{g} \hookleftarrow \ldots \hookleftarrow \{0\}.$$

Similarly, the *ascending central series* $(\mathfrak{C}_p\mathfrak{g})_{p \geqq 0}$ of $\mathfrak{g}$ is defined recursively by setting

$$\begin{cases} \mathfrak{C}_0\mathfrak{g} = \{0\}, \\ \mathfrak{C}_{p+1}\mathfrak{g} = \text{preimage of the centre of } \mathfrak{g}/\mathfrak{C}_p\mathfrak{g} \text{ under the canonical} \\ \qquad\qquad \text{epimorphism } \mathfrak{g} \to \mathfrak{g}/\mathfrak{C}_p\mathfrak{g}. \end{cases}$$

Then the ideal $\mathfrak{c} = \mathfrak{C}_1\mathfrak{g}$ of $\mathfrak{g}$ is the centre of $\mathfrak{g}$ and the following ascending filtration of *ideals* of $\mathfrak{g}$ arises:

$$\{0\} \hookrightarrow \mathfrak{C}_1\mathfrak{g} \hookrightarrow \mathfrak{C}_2\mathfrak{g} \hookrightarrow \ldots \hookrightarrow \mathfrak{C}_p\mathfrak{g} \hookrightarrow \ldots \hookrightarrow \mathfrak{g},$$

where

$$\mathfrak{C}_{p+1}\mathfrak{g} = \{X \in \mathfrak{g} \mid [X,Y] \in \mathfrak{C}_p\mathfrak{g} \textit{ for all } Y \in \mathfrak{g}\}.$$

4.8 Theorem. Let $\mathfrak{g}$ be a Lie algebra over k and $p_0 \geqq 1$ a sufficiently large natural number. The following conditions are pairwise equivalent:

(i) $\mathfrak{C}^{p_0}\mathfrak{g} = \{0\}$,

(ii) In $\mathfrak{g}$ there exists a sequence $(\mathfrak{g}_p)_{0 \leqq p \leqq p_0}$ of ideals such that

$\mathfrak{g}_0 = \mathfrak{g}$, $\mathfrak{g}_{p_0} = \{0\}$, and

$$\mathfrak{g}_0 \hookleftarrow \mathfrak{g}_1 \hookleftarrow \mathfrak{g}_2 \hookleftarrow \ldots \hookleftarrow \mathfrak{g}_p \hookleftarrow \ldots \hookleftarrow \mathfrak{g}_{p_0},$$

and

$$[\mathfrak{g},\mathfrak{g}_p] \subseteqq \mathfrak{g}_{p+1} \qquad (0 \leqq p \leqq p_0-1),$$

(iii) $\mathfrak{C}_{p_0}\mathfrak{g} = \mathfrak{g}$,

(iv) $\mathrm{ad}_{\mathfrak{g}}X_1 \circ \mathrm{ad}_{\mathfrak{g}}X_2 \circ \ldots \circ \mathrm{ad}_{\mathfrak{g}}X_{p_0} = 0$ for all elements $(X_p)_{0 \leqq p \leqq p_0}$ in $\mathfrak{g}$.

Proof. The implications (i) ⇒ (ii) ⇒ (iii) ⇒ (i) follow as in Theorem 4.4. The equivalence of (i) and (iv) follows from the fact that $\mathfrak{C}^p\mathfrak{g}$ is the set of all linear combinations of elements of $\mathfrak{g}$ having the form

$$[X_1,[X_2,\ldots,[X_{p-2},[X_{p-1},X_p]] \ldots]]$$

where $X_1,X_2,\ldots,X_p$ are arbitrary elements of $\mathfrak{g}$. —

The Lie algebra $\mathfrak{g}$ over k is said to be *nilpotent*, if $\mathfrak{g}$ satisfies one hence all of the conditions of Theorem 4.8. If $p_0 \geqq 1$ and $\mathfrak{C}^{p_0-1}\mathfrak{g} \neq \{0\}$ - then p_0 is called to be the *length* of the nilpotent Lie algebra $\mathfrak{g}$. Moreover, (iv) is called to be the *local nilpotency condition* of length p_0 satisfied by $\mathfrak{g}$.

Corollary 1. Let $\mathfrak{g}$ be a nilpotent Lie algebra over k.

(i) Each Lie subalgebra $\mathfrak{h}$ of $\mathfrak{g}$ is nilpotent.

(ii) For each ideal $\mathfrak{h}$ of $\mathfrak{g}$ the quotient Lie algebra $\mathfrak{g}/\mathfrak{h}$ is nilpotent.

Proof. Let p_0 denote the length of $\mathfrak{g}$. In the case (i) we have $\mathfrak{C}^{p_0}\mathfrak{h} \subseteqq \mathfrak{C}^{p_0}\mathfrak{g} = \{0\}$. In the case (ii) let $\mathfrak{k} = \mathfrak{g}/\mathfrak{h}$ and $\pi:\mathfrak{g} \to \mathfrak{k}$ be the canonical epimorphism - then $\mathfrak{C}^{p_0}\mathfrak{k} = \pi(\mathfrak{C}^{p_0}\mathfrak{g}) = \{0\}$. —

Corollary 2. Let $\mathfrak{g} \neq \{0\}$ be a nilpotent Lie algebra over k with centre $\mathfrak{c}$ - then $\mathfrak{c} \neq \{0\}$.

Corollary 3. Let K be a commutative field and $\mathfrak{g}$ a Lie algebra over K with $\dim_K\mathfrak{g} = n$. Then $\mathfrak{g}$ is nilpotent if and only if there exists a sequence $(\mathfrak{g}_j)_{0 \leqq j \leqq n}$ of ideals of $\mathfrak{g}$ such that $\mathfrak{g}_0 = \mathfrak{g}$, $\mathfrak{g}_n = \{0\}$, and

$$\mathfrak{g}_0 \hookleftarrow \mathfrak{g}_1 \hookleftarrow \ldots \hookleftarrow \mathfrak{g}_n,$$

and such that

$$[\mathfrak{g},\mathfrak{g}_j] \subseteqq \mathfrak{g}_{j+1},$$

$$\dim_K\mathfrak{g}_j = n-j$$

holds for $0 \leqq j \leqq n-1$.

Proof. In view of Theorem 4.8 it will be sufficient to prove that for a nilpotent Lie algebra $\mathfrak{g}$ over K the decreasing sequence $(\mathfrak{g}_j)_{0 \leqq j \leqq n}$ of ideals of $\mathfrak{g}$ can be constructed in such a way that $[\mathfrak{g},\mathfrak{g}_j] \subseteqq \mathfrak{g}_{j+1}$ and $\dim_K \mathfrak{g}_j = n-j$ holds for $0 \leqq j \leqq n-1$. Let $p \geqq 0$ be a natural number such that $\mathfrak{C}^p\mathfrak{g} = \{0\}$ and let $r \geqq 1$ be the codimension of $\mathfrak{C}^{p+1}\mathfrak{g}$ in $\mathfrak{C}^p\mathfrak{g}$. There exists a sequence $(\mathfrak{h}_j)_{0 \leqq j \leqq r}$ such that $\mathfrak{h}_0 = \mathfrak{C}^p\mathfrak{g}$, $\mathfrak{h}_r = \mathfrak{C}^{p+1}\mathfrak{g}$,

$$\mathfrak{h}_0 \longleftarrow \mathfrak{h}_1 \longleftarrow \ldots \longleftarrow \mathfrak{h}_r$$

and

$$\dim_K \mathfrak{h}_j/\mathfrak{h}_{j+1} = 1$$

for $0 \leqq j \leqq r-1$. The vector subspaces $(\mathfrak{h}_j)_{0 \leqq j \leqq r}$ are ideals of $\mathfrak{g}$ in view of the inclusions

$$[\mathfrak{g},\mathfrak{h}_j] \subseteqq [\mathfrak{g},\mathfrak{C}^p\mathfrak{g}] = \mathfrak{C}^{p+1}\mathfrak{g} \subseteqq \mathfrak{h}_{j+1} \quad (0 \leqq j \leqq r-1).$$

Thus the decreasing sequence $(\mathfrak{h}_j)_{0 \leqq j \leqq r}$ of ideals with codimension 1 in the preceding ideal *refines* the step from $\mathfrak{C}^p\mathfrak{g}$ to $\mathfrak{C}^{p+1}\mathfrak{g}$ in the correct way. By applying this refinement procedure for every natural number $p \geqq 0$ such that $\mathfrak{C}^p\mathfrak{g} \neq \{0\}$, the statement follows. —

Remarks.

1. A Lie algebra $\mathfrak{g}$ over the commutative ring k is called to be *abelian* if any two of its elements are commuting. Thus the one-step nilpotent Lie algebras $\mathfrak{g}$ over k are formed by the abelian Lie algebras $\mathfrak{g} \neq \{0\}$ over k.

2. The *derived series* $(\mathfrak{D}^p\mathfrak{g})_{p \geqq 0}$ of the Lie algebra $\mathfrak{g}$ over k is defined recursively via the prescription

$$\begin{cases} \mathfrak{D}^0\mathfrak{g} = \mathfrak{g}, \\ \mathfrak{D}^{p+1}\mathfrak{g} = [\mathfrak{D}^p\mathfrak{g}, \mathfrak{D}^p\mathfrak{g}]. \end{cases}$$

The Lie algebra $\mathfrak{g}$ is called to be *solvable* if the descending filtration $(\mathfrak{D}^p\mathfrak{g})_{p \geqq 0}$ of *ideals* of $\mathfrak{g}$ terminates at $\{0\}$, i.e., if we have

$$\mathfrak{D}^{p_0}\mathfrak{g} = \{0\}$$

for a sufficiently large natural number $p_0 \geqq 1$. Every nilpotent Lie algebra

$\mathfrak{g}$ is solvable, the converse being false.

3. It follows from Theorem 4.8 that for any nilpotent Lie algebra $\mathfrak{g}$ over a commutative ring k and for each element $X \in \mathfrak{g}$ the adjoint linear mapping $ad_{\mathfrak{g}}X$ of X (Section 4.7) is *nilpotent*, i.e., there is a sufficiently large natural number $p_0 \geqq 1$ so that

$$(ad_{\mathfrak{g}}X)^{p_0} = 0.$$

If K is a commutative field and $\mathfrak{g}$ is of finite dimension over K, the converse also is true, i.e., the nilpotency of the adjoint linear mappings $ad_{\mathfrak{g}}X$ for all $X \in \mathfrak{g}$ implies that $\mathfrak{g}$ is a nilpotent Lie algebra over K.

4.9 Let G denote a *real Lie group*. Recall that the underlying locally compact topological space is a real infinitely differentiable manifold that is countable at infinity. Moreover, it should be observed that every Lie group G can be equipped with the structure of an analytic Lie group that is compatible with its infinitely differentiable structure. The linear differential operators D with real infinitely differentiable coefficients on G form an associative algebra over the field $\mathbb{R}$ and therefore a Lie algebra over $\mathbb{R}$ with respect to the bracket operation

$$(D_1,D_2) \rightsquigarrow [D_1,D_2] = D_1 \circ D_2 - D_2 \circ D_1.$$

The bracket of two infinitely differentiable vector fields on G is again a vector field on G of this kind. It follows that the infinitely differentiable vector fields on G form a Lie subalgebra of the real Lie algebra of linear differential operators on G mentioned above. Among the infinitely differentiable vector fields on G we may single out the *left invariant* vector fields X on G which form a Lie subalgebra

$$\mathfrak{g} = \mathrm{Lie}(G).$$

The Lie algebra $\mathfrak{g}$ over $\mathbb{R}$ is called to be *the Lie algebra* of the real Lie group G. If $T(G,1_G)$ denotes the *tangent space* to G at the neutral element 1_G of G, the mapping which assigns to each vector field $X \in \mathfrak{g}$ its value at the point $1_G \in G$ is an isomorphism of the real vector space $\mathfrak{g}$ onto the real vector space $T(G,1_G)$ which allows to identify $\mathfrak{g}$ with $T(G,1_G)$. Let G' be another real Lie group with neutral element $1_{G'} \in G'$ and Lie algebra

$\mathfrak{g}' = \mathrm{Lie}(G') = T(G',1_{G'})$. Each analytic group morphism $\Phi: G \to G'$ defines by its *differential* a $\mathbb{R}$-linear mapping $\Phi': T(G,1_G) \to T(G',1_{G'})$ which is a Lie algebra morphism of $\mathfrak{g}$ into $\mathfrak{g}'$, i.e., the identity

$$\Phi'([X,Y]) = [\Phi'(X), \Phi'(Y)]$$

holds for all pairs $(X,Y) \in \mathfrak{g} \times \mathfrak{g}$.

The simplest example of a real Lie group is the one-dimensional compact *torus group* $\mathbb{T} = \mathbb{R}/\mathbb{Z}$. The Lie group $\mathbb{T}$ under its additive group law may be identified with the compact Lie group $G = \underline{\underline{U}}(1,\mathbb{C}) = \{z \in \mathbb{C} \mid |z| = 1\}$ under its multiplicative law. The tangent space $\mathfrak{g}$ at the identity $1_{\underline{\underline{U}}(1,\mathbb{C})}$ is $\mathfrak{g} = 2\pi i\mathbb{R}$.

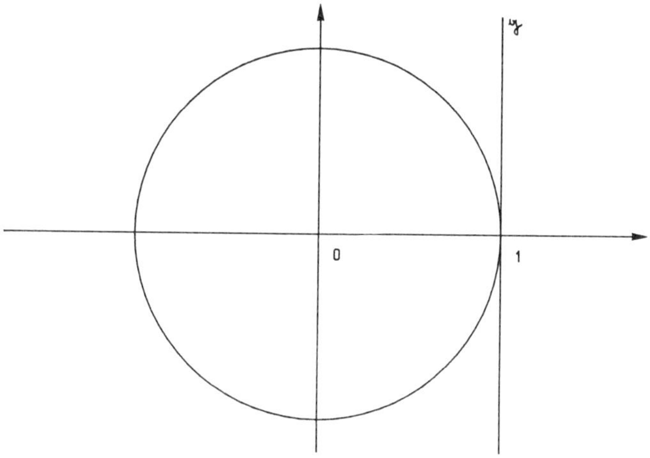

The "basic" example of a real Lie group is the *full linear group* $\underline{\underline{GL}}(n,\mathbb{R})$ of *invertible* $n \times n$ matrices with real entries. In this case the Lie algebra $\mathfrak{gl}(n,\mathbb{R}) = \mathrm{Lie}(\underline{\underline{GL}}(n,\mathbb{R}))$ consists of *all* real $n \times n$ matrices and the Lie algebra structure on $\mathfrak{gl}(n,\mathbb{R})$ is given by the commutator

$$[X,Y] = XY - YX$$

for all pairs $(X,Y) \in \mathfrak{gl}(n,\mathbb{R}) \times \mathfrak{gl}(n,\mathbb{R})$. For the present purpose, it will be sufficient to consider *linear Lie groups*. A linear Lie group G is a *closed* subgroup of the full linear group $\underline{\underline{GL}}(n,\mathbb{R})$ for a suitably chosen integer $n \geq 1$. Its Lie algebra $\mathfrak{g} = \mathrm{Lie}(G)$ then is a Lie subalgebra of $\mathfrak{gl}(n,\mathbb{R})$:

$$\underline{\underline{G}} \subseteq \underline{\underline{GL}}(n,\mathbb{R}), \quad \mathfrak{g} \subseteq \mathfrak{gl}(n,\mathbb{R})$$

The corresponding notion of bracket of $\mathfrak{g}$ is therefore the *restriction* of the preceding operation to $\mathfrak{g}$. i.e., $[X,Y] = XY - YX$ for X,Y in $\mathfrak{g}$.

4.10 Lemma. Let G be a real Lie group and $(\mathfrak{C}_p G)_{p \geq 0}$ the ascending central series of G. Then $\mathfrak{C}_p G$ is a closed normal Lie subgroup of G for each natural number p and we have

$$\mathrm{Lie}(\mathfrak{C}_p G) = \mathfrak{C}_p \mathrm{Lie}(G) \quad (p \geq 0).$$

Proof. Let $\mathfrak{g} = \mathrm{Lie}(G)$ be the Lie algebra of G. In the case $p = 0$ we have $\mathfrak{C}_0 G = \{1_G\}$ and $\mathfrak{C}_0 \mathfrak{g} = \{0\}$. Moreover, for $p = 1$ the centre $C = \mathfrak{C}_1 G$ is closed in G and its Lie algebra $\mathrm{Lie}(\mathfrak{C}_1 G)$ coincides with the centre $\mathfrak{c} = \mathfrak{C}_1 \mathfrak{g}$ of $\mathfrak{g}$. Suppose that there is a natural number $p \geq 1$ such that the identity above is true. Let $G' = G/\mathfrak{C}_p G$ and C' the centre of G'. Denote by $\pi: G \to G'$ and $\pi': G' \to G'/C'$ the canaonical epimorphisms. Then we have by Section 4.2 supra

$$\mathfrak{C}_{p+1} G = \mathrm{Ker}(\pi' \circ \pi).$$

Consequently $\mathfrak{C}_{p+1} G$ is a closed normal Lie subgroup of G. Moreover $\mathrm{Lie}(\mathfrak{C}_{p+1} G)$ is the kernel of the differential of the Lie group morphism $\pi' \circ \pi: G \to G'/C'$ and therefore coincides with $\mathfrak{C}_{p+1} \mathfrak{g}$. This proves the lemma by induction on $p \geq 0$. —

Remark. For any connected real Lie group G the descending central series $(\mathfrak{C}^p G)_{p \geq 0}$ of G is formed by the (not necessarily closed) connected normal Lie subgroups $\mathfrak{C}^p G$ of G and we have

$$\mathrm{Lie}(\mathfrak{C}^p G) = \mathfrak{C}^p \mathrm{Lie}(G)$$

for each natural number p.

4.11 Theorem. Let G denote a real Lie group - then the following conditions are equivalent:

(i) The neutral component G_1 of G is a nilpotent group.

(ii) The Lie algebra $\mathfrak{g} = \mathrm{Lie}(G)$ of G is a nilpotent Lie algebra over the field $\mathbb{R}$.

Proof. (i) ⇒ (ii): The neutral component G_1 of G is a closed normal Lie subgroup of G. Let G_1 be a nilpotent group of length $p_1 \geq 1$. Then $\mathfrak{C}_{p_1} G_1 = G_1$

by Theorem 4.4. In view of Lemma 4.10 we obtain $\mathfrak{C}_{p_1}\mathfrak{g} = \mathfrak{g}$. An application of Theorem 4.8 yields the result.

(ii) ⇒ (i): Let the nilpotent Lie algebra $\mathfrak{g}$ over $\mathbb{R}$ have length $p_1 \geqq 1$ - then $\mathfrak{C}_{p_1}\mathfrak{g} = \mathfrak{g}$ by Theorem 4.8. It follows from Lemma 4.10 that $\mathfrak{C}_{p_1}G_1$ is an open and closed normal Lie subgroup of G_1. Hence $\mathfrak{C}_{p_1}G_1 = G_1$, i.e., G_1 is a nilpotent group by Theorem 4.4. ——

For any real Lie group G with Lie algebra $\mathfrak{g} = \mathrm{Lie}(G)$ we denote by

$$\exp_G : \mathfrak{g} \to G$$

the *exponential mapping*. Then for any analytic group morphism Φ of G into the real Lie group G' with Lie algebra $\mathfrak{g}'$ the following diagram is commutative:

$$\begin{array}{ccc} \mathfrak{g} & \xrightarrow{\Phi'} & \mathfrak{g}' \\ {\scriptstyle \exp_G}\downarrow & & \downarrow{\scriptstyle \exp_{G'}} \\ G & \xrightarrow[\Phi]{} & G' \end{array}$$

For all $x \in G$ denote by

$$\mathrm{Ad}_{\mathfrak{g}}(x) : \mathfrak{g} \to \mathfrak{g}$$

the automorphism of $\mathfrak{g}$ which is the differential of the inner automorphism $\mathrm{Int}_G(x)$ of G associated with the element x (Section 2.11). The action of G on $\mathfrak{g}$ by $\mathrm{Ad}_{\mathfrak{g}}$ is called the *adjoint action* of G. It follows that $\exp_G$ is a *local* isomorphism of $\mathfrak{g}$ into G which makes the following diagram commutative for all $x \in G$:

$$\begin{array}{ccc} \mathfrak{g} & \xrightarrow{\mathrm{Ad}_{\mathfrak{g}}(x)} & \mathfrak{g} \\ {\scriptstyle \exp_G}\downarrow & & \downarrow{\scriptstyle \exp_G} \\ G & \xrightarrow[\mathrm{Int}_G(x)]{} & G \end{array}$$

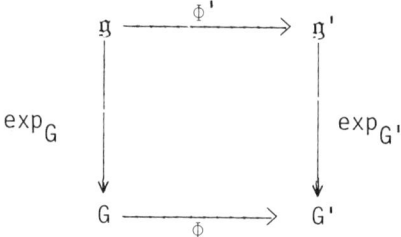

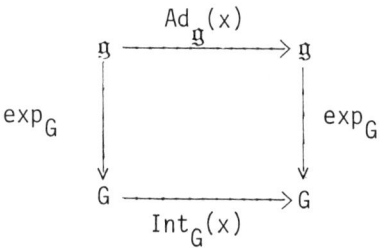

For instance, in the simplest case $G = \mathbb{T}$, the exponential mapping $\exp_{\mathbb{T}}$ is the map $\mathfrak{g} \ni 2\pi i\theta \rightsquigarrow e^{2\pi i\theta} \in \mathbb{T}$ $(\theta \in \mathbb{R})$ and the adjoint action $Ad_{\mathfrak{g}}$ of G on $\mathfrak{g} = 2\pi i\mathbb{R}$ is trivial.

In the basic example of a real Lie group $\underline{\underline{GL}}(n,\mathbb{R})$ with Lie algebra $\mathfrak{gl}(n,\mathbb{R})$ as mentioned in Section 4.9 supra, the exponential mapping $\exp_{\underline{\underline{GL}}(n,\mathbb{R})}$ is the usual matrix exponential

$$\exp_{\underline{\underline{GL}}(n,\mathbb{R})} X = \sum_{\nu \geq 0} \frac{1}{\nu!} X^{\nu} \qquad (X \in \mathfrak{gl}(n,\mathbb{R})).$$

The adjoint action of $\underline{\underline{GL}}(n,\mathbb{R})$ is given by conjugation

$$Ad_{\mathfrak{gl}(n,\mathbb{R})}(x)X = Int_{\underline{\underline{GL}}(n,R)}(x)X = xXx^{-1}$$

for $x \in \underline{\underline{GL}}(n,\mathbb{R})$ and $X \in \mathfrak{gl}(n,\mathbb{R})$. For a linear Lie group $G \subseteq \underline{\underline{GL}}(n,\mathbb{R})$ with Lie algebra $\mathfrak{g} \subseteq \mathfrak{gl}(n,\mathbb{R})$, the corresponding notion of exponential mapping $\exp_G$ and adjoint action $Ad_{\mathfrak{g}}$ of G are therefore the restrictions to $\mathfrak{g}$ and G of the preceding operations, i.e., we still have

$$\exp_G X = \sum_{\nu \geq 0} \frac{1}{\nu!} X^{\nu} \qquad (X \in \mathfrak{g}),$$

and

$$Ad_{\mathfrak{g}}(x)X = Int_G(x)X = xXx^{-1}$$

for $X \in G$ and $X \in \mathfrak{g}$.

For an arbitrary real Lie group G with Lie algebra $\mathfrak{g} = Lie(G)$ the image $\exp_G(\mathfrak{g})$ of $\mathfrak{g}$ is contained in the neutral component G_1 of G. However, $\exp_G$ does not need to be a surjective mapping. If G is a simply connected Lie group and $\exp_G$ is a *global* diffeomorphism of $\mathfrak{g} = Lie(G)$ *onto* G then G is called to be *exponential*.

4.12 Theorem. The simply connected, nilpotent, real Lie groups G are exponential.

Proof. Let $\mathfrak{g} = Lie(G)$ be the Lie algebra of G. We shall proceed by induction on $\dim_{\mathbb{R}}\mathfrak{g} \geq 1$. If $\dim_{\mathbb{R}}\mathfrak{g} = 1$, the result is clear. Suppose that the theorem holds for $1 \leq \dim_{\mathbb{R}}\mathfrak{g} \leq n$ and assume $\dim_{\mathbb{R}}\mathfrak{g} = n+1$. Let $\mathfrak{c} = \mathfrak{C}_1\mathfrak{g}$ be the centre of $\mathfrak{g}$. By virtue of Corollary 2 of Theorem 4.8 we have $\mathfrak{c} \neq \{0\}$. Let

$$C = \mathfrak{C}_1 G = \exp_G \mathfrak{c}$$

be the centre of G. By Corollary 1 of Theorem 4.4 the quotient G/C is a simply connected, nilpotent, real Lie group. The inductive hypothesis implies that $\exp_{G/C}:\mathfrak{g}/\mathfrak{c} \to G/C$ is a global diffeomorphism of $\mathfrak{g}/\mathfrak{c}$ onto G/C.

Let $\pi:G \to G/C$ denote the canonical epimorphism and let also $\pi:\mathfrak{g} \to \mathfrak{g}/\mathfrak{c}$ be its differential. Then we have the identity

$$\pi(\exp_G X) = \exp_{G/C}\pi(X)$$

for all $X \in \mathfrak{g}$. Choose a vector subspace M of $\mathfrak{g}$ such that $\mathfrak{g} = M \oplus \mathfrak{c}$ holds. It follows that the restriction $\exp_G|M$ defines a diffeomorphism of M onto its image $\exp_G(M)$.

Suppose that $X_j \in M$ and $Z_j \in \mathfrak{c}$ ($j \in \{1,2\}$) satisfy the identity

$$\exp_G(X_1 + Z_1) = \exp_G(X_2 + Z_2).$$

By virtue of $[X_j,Z_j] = 0$ we have $\exp_G(X_j+Z_j) = \exp_G X_j.\exp_G Z_j$ for $j \in \{1,2\}$. Hence the identity

$$\pi(\exp_G X_1) = \pi(\exp_G X_2)$$

and therefore $X_1 = X_2$. Since $\pi(\exp_G M) = G/C$ we see that $G = (\exp_G(M)).\ C$. Thus $\exp_G$ is a surjective mapping of $\mathfrak{g}$ onto G. It follows that the mapping defined via the prescription

$$M \times C \ni (X,z) \quad > (\exp_G X).z \in G/C$$

is bijective and open, hence a diffeomorphism. Since M is simply connected, this implies that C is simply connected. But then $\exp_G:\mathfrak{g} \to G$ is a diffeomorphism of $\mathfrak{g}$ onto G. —

4.13 Lemma. Let G be a simply connected real Lie group and H a connected normal Lie subgroup of G. Then H is closed in G and the quotient group G/H is a simply connected real Lie group.

Proof. Let $\mathfrak{g} = \mathrm{Lie}(G)$ and $\mathfrak{h} = \mathrm{Lie}(H)$ be the Lie algebras of G and H, respectively. Then $\mathfrak{h}$ is an ideal in $\mathfrak{g}$. Form the quotient Lie algebra $\mathfrak{g}' = \mathfrak{g}/\mathfrak{h}$ and let G' be the simply connected real Lie group, such that $\mathfrak{g}' = \mathrm{Lie}(G')$. Let $\Phi:G \to G'$ be the Lie group morphism such that its differential is the canonical epimorphism $\phi:\mathfrak{g} \to \mathfrak{g}'$. Then $H' = \mathrm{Ker}(\Phi)$ is a closed subgroup of G and its neutral component H_1' is also closed in G. In view of $\mathfrak{h} = \mathrm{Lie}(H')$

we have $H = H_1'$. Hence H is a closed subgroup of G. The Lie group morphism $\Phi: G \to G'$ gives rise to a morphism

$$\tilde{\Phi} : G/H \to G'.$$

Since $\tilde{\Phi}$ is a covering map and G' is simply connected, we conclude that Φ is an isomorphism of G/H onto G'. It follows that G/H is a simply connected real Lie group. ——

4.14 Theorem. The simply connected, nilpotent, real Lie groups G, the connected Lie subgroups H of G, and the quotient Lie groups G/H where H denotes a connected normal Lie subgroup of G, are unimodular, locally compact, topological groups.

Proof. Let $\mathfrak{g}$ = Lie(G) be the Lie algebra of G. Then for all $x \in G$ and all $X \in \mathfrak{g}$ the identities

$$\mathrm{Int}_G(x)\ (\exp_G X) = x(\exp_G X)x^{-1} = \exp_G(\mathrm{Ad}_G(x)X)$$

hold. Therefore we have

$$\gamma_G \times \delta_G(x)(\exp_G X) = \exp_G(\mathrm{Ad}_G(x)X).$$

Thus the right modular function Δ_G of G (cf. Section 2.3) takes the form

$$\Delta_G(x) = |\det \mathrm{Ad}_G(x^{-1})|$$

for $x \in G$. Since $\mathrm{ad}_{\mathfrak{g}} X$ is a nilpotent endomorphism of $\mathfrak{g}$ for all $X \in \mathfrak{g}$ by Section 4.8, Remark 3, we have

$$\mathrm{tr}(\mathrm{ad}_{\mathfrak{g}} X) = 0$$

and therefore

$$\det \mathrm{Ad}_G(x) = 1$$

for all $x \in G$. Hence $\Delta_G = 1$, i.e., G is unimodular. Since each connected subgroup H of G is a simply connected closed Lie subgroup of G, the real Lie group H is also unimodular. Finally, if H denotes a connected normal Lie subgroup of G, the real Lie group G/H is simply connected by Lemma 4.13. —

It follows by an induction argument that for simply connected, nilpotent, real Lie groups G the image of Lebesgue measure dX of $\mathfrak{g}$ = Lie(G) under the diffeomorphism $\exp_G:\mathfrak{g} \to G$ is a left and right Haar measure dx on G. Thus we have

$$\int_G \phi(x)dx = \int_{\mathfrak{g}} \phi(\exp_G X)dX$$

for all functions $\phi \in \mathbb{K}(G)$.

4.15 Let G denote a simply connected, nilpotent, real Lie group with centre $C = \mathfrak{C}_1(G)$. Then C is a simply connected Lie subgroup of G. Let $(U,\mathcal{H})$ denote a topologically irreducible, continuous, unitary, linear representation of G with central character χ_U (Section 1.6). If the dimension of C is $\geqq 2$ then the kernel G' of χ_U is a connected closed normal Lie subgroup of G of dimension $\geqq 1$ which is contained in the kernel of U. Set G" = G/G' and let $\pi:G \to G''$ denote the canonical epimorphism. There exists a topologically irreducible, continuous, unitary, linear representation $(V,\mathcal{H})$ of G" such that

$$U = V \circ \pi$$

holds and G" is by Corollary 1 of Theorem 4.4 and Lemma 4.13 a simply connected, nilpotent, real Lie group. In other words: In the case that the dimension of C is $\geqq 2$ each topologically irreducible, continuous, unitary, linear representation $(U,\mathcal{H})$ of G gives rise to a topologically irreducible, continuous, unitary, linear representation $(V,\mathcal{H})$ of G" where G" is a connected closed normal Lie subgroup of G of dimension $\geqq 1$.

4.16 Now assume that the simply connected, nilpotent, real Lie group G has dimension > 1 and that its centre C is one-dimensional. Then the centre $\mathfrak{c} = \mathfrak{C}_1\mathfrak{g}$ of $\mathfrak{g}$ = Lie(G) satisfies $\dim_{\mathbb{R}}\mathfrak{c} = 1$ and we have $\mathfrak{c} \subsetneqq \mathfrak{C}_2\mathfrak{g}$ in the ascending central series $(\mathfrak{C}_p\mathfrak{g})_{p \geqq 0}$ of $\mathfrak{g}$. Since

$$\mathfrak{C}_2\mathfrak{g} = \{Y \in \mathfrak{g} \mid [X,Y] \in \mathfrak{c} \text{ for all } X \in \mathfrak{g}\},$$

there exists a vector field $Y_0 \in \mathfrak{C}_2\mathfrak{g}$ such that $Y_0 \notin \mathfrak{c}$. We have

$$[\mathfrak{g},Y_0] \subseteqq \mathfrak{c}.$$

Let $T_0 \neq 0$ be an element of $\mathfrak{c}$. Then $\mathfrak{c} = \mathbb{R}.T_0$ and there exists a $\mathbb{R}$-linear form λ_0 on the real vector space $\mathfrak{g}$ such that

$$[X,Y_0] = \lambda_0(X)T_0$$

holds for all elements $X \in \mathfrak{g}$. Denote by $\mathfrak{g} \times \mathfrak{g}^* \ni (X,\lambda) \rightsquigarrow \langle X,\lambda\rangle \in \mathbb{R}$ the canonical bilinear form associated with the real vector space $\mathfrak{g}$ and its dual vector space $\mathfrak{g}^*$ over $\mathbb{R}$. In this notation we have $\lambda_0(X) = \langle X,\lambda_0\rangle$ for all $X \in \mathfrak{g}$. Since $\lambda_0 \neq 0$ by virtue of $Y_0 \notin \mathfrak{c}$, there exists a vector field $X_0 \in \mathfrak{g}$ such that $\langle X_0,\lambda_0\rangle = 1$. Let $\mathfrak{g}_0 = \mathrm{Ker}(\lambda_0) = \mathrm{Ker}(\mathrm{ad}_{\mathfrak{g}} Y_0)$ be the *centralizer* of Y_0 in $\mathfrak{g}$. Then $\mathfrak{g}_0$ is a vector subspace of $\mathfrak{g}$ and, moreover, an *ideal* of $\mathfrak{g}$ in view of the identity

$$\langle [X,Y],\lambda_0\rangle = 0$$

which holds for $X \in \mathfrak{g}_0$, $Y \in$. Obviously we have the ascending filtration of strict inclusions

$$\mathfrak{c} \hookrightarrow \mathfrak{g}_0 \hookrightarrow \mathfrak{g}$$

where $\mathfrak{g}_0$ has real codimension 1 in $\mathfrak{g}$, and the direct sum decomposition

$$\mathfrak{g} = \mathbb{R}X_0 \oplus \mathfrak{g}_0$$

holds. If G_0 denotes the (unique) connected Lie subgroup of G such that $\mathfrak{g}_0 = \mathrm{Lie}(G_0)$ - then we have the ascending filtration of strict inclusions

$$C \hookrightarrow G_0 \hookrightarrow G$$

and G is the semi-direct product of the simply connected, normal Lie subgroup G_0 and the one-parameter subgroup $\exp_G(\mathbb{R}X_0)$ generated by $X_0 \in \mathfrak{g}$.

4.17 Keep to the preceding notations and conventions. Define

$$\mathfrak{n}_0 = \mathbb{R}Y_0 \oplus \mathbb{R}T_0.$$

Obviously $\mathfrak{n}_0$ is an abelian ideal in $\mathfrak{g}$ of real dimension 2 containing $\mathfrak{c}$ and such that $[\mathfrak{n}_0,\mathfrak{g}] \subseteq \mathfrak{c}$. Moreover, the ideal $\mathfrak{g}_0$ in $\mathfrak{g}$ is the centralizer of $\mathfrak{n}_0$ in $\mathfrak{g}$. The unique connected Lie subgroup N_0 of G such that $\mathfrak{n}_0 = \mathrm{Lie}(N_0)$ is a closed, normal, abelian subgroup of G. We will establish (Theorem 4.19 infra) that N_0 is *properly embedded into* G (Section 2.19), i.e., that for each unitary character $\hat{y} \in \hat{N}_0$ its orbit $G.\hat{y}$ under the left action of G on $\hat{N}_0$ by inner automorphisms of N_0 is a locally compact, topological subspace of the locally compact, topological group $\hat{N}_0$. For all elements $x \in G$ and all vector fields $X \in \mathfrak{n}_0$ we obtain (cf. Section 4.14)

$$\begin{aligned} x.\hat{y}(\exp_{N_0} X) &= \hat{y} \circ \operatorname{Int}_{N_0}(x^{-1})(\exp_{N_0} X) \\ &= \hat{y}(x^{-1}(\exp_{N_0} X)x) \qquad\qquad (\hat{y} \in \hat{N}_0) \\ &= \hat{y}(\exp_{N_0}(\operatorname{Ad}_G(x^{-1})X)) \end{aligned}$$

where $\exp_{N_0} = \exp_G | N_0$.

The next step is to observe that the Pontryagin dual $\hat{N}_0$ of N_0 is topologically isomorphic to the dual vector space $\mathfrak{n}_0^*$ of the two-dimensional, locally compact, topological vector space $\mathfrak{n}_0$ over $\mathbb{R}$ with basis $\{Y_0, T_0\}$. Indeed, since N_0 is a simply connected, abelian, real Lie group there exists for each unitary character $\hat{y} \in \hat{N}_0$ a unique $\mathbb{R}$-linear form $\ell \in \mathfrak{n}_0^*$ such that

$$\hat{y}(\exp_{N_0} X) = e^{2\pi i \langle X, \ell \rangle}$$

for all $X \in \mathfrak{n}_0$ and the assignment

$$\hat{N}_0 \ni \hat{y} \rightsquigarrow \ell \in \mathfrak{n}_0^*$$

furnishes the desired topological isomorphism of $\hat{N}_0$ onto $\mathfrak{n}_0^*$. If we introduce the $\mathbb{R}$-linear mapping

$$\operatorname{CoAd}_G(x) : \mathfrak{g}^* \to \mathfrak{g}^*$$

which is the *transpose* of the $\mathbb{R}$-linear mapping

$$\operatorname{Ad}_G(x^{-1}) : \mathfrak{g} \to \mathfrak{g}$$

for all elements $x \in G$, then we get the identity

$$x.\hat{y}(\exp_{N_0} X) = e^{2\pi i \langle X, \operatorname{CoAd}_G(x)\ell \rangle}$$

for all $X \in \mathfrak{n}_0$. The preceding formula allows *to compute explicitly the orbit* $G.\hat{y}$ *of each unitary character* $\hat{y} \in \hat{N}_0$ *under the left action of* G *on* $\hat{N}_0$ *by inner automorphisms of* N_0 *using the coadjoint orbit* $\operatorname{CoAd}_G(G)$ *in* $\mathfrak{n}_0^*$ *of the* *linear form* $\ell \in \mathfrak{n}_0^*$ *associated with* $\hat{y}$.

4.18 In Section 4.16 we established the fact that G is the semi-direct product of the simply connected Lie subgroup G_0 and the one-parameter subgroup $\exp_G(\mathbb{R}X_0)$ generated by $X_0 \in \mathfrak{g}$.

For each element $x \in G_0$ there exists $X \in \mathfrak{g}_0$ such that $x = \exp_G X$ holds. Therefore we have

$$\mathrm{Ad}_G(x) = \exp_G(\mathrm{ad}_{\mathfrak{g}} X) = \sum_{n \geqq 0} \frac{1}{n!} (\mathrm{ad}_{\mathfrak{g}} X)^n.$$

Since $\mathrm{ad}_{\mathfrak{g}} X | \mathfrak{n}_0 = 0$ we have

$$\mathrm{CoAd}_G(x) | \mathfrak{n}_0^* = \mathrm{id}_{\mathfrak{n}_0^*}$$

for all $x \in G_0$. Thus the coadjoint action of G_0 on $\mathfrak{n}_0^*$ is trivial.

Furthermore, if $Y \in \mathfrak{n}_0$ there exists a pair $(\kappa,\lambda) \in \mathbb{R}^2$ such that $Y = \kappa Y_0 + \lambda T_0$. For all $t \in \mathbb{R}$ we obtain

$$\begin{aligned} \mathrm{Ad}_G(\exp_G(tX_0))(Y) &= \exp_G(t.\mathrm{ad}_{\mathfrak{g}} X_0)(\kappa Y_0 + \lambda T_0) \\ &= (\kappa Y_0 + \lambda T_0) + t\kappa T_0 \\ &= \kappa Y_0 + (\kappa t + \lambda) T_0. \end{aligned}$$

Let $\{Y_0^*, T_0^*\}$ denote the basis of $\mathfrak{n}_0^*$ which is dual to the basis $\{Y_0, T_0\}$ of $\mathfrak{n}_0$. It follows that the endomorphism $\mathrm{Coad}_G(\exp_G(tX_0)) | \mathfrak{n}_0^*$ of the two-dimensional real vector space $\mathfrak{n}_0^*$ admits for all $t \in \mathbb{R}$ the matrix

$$\begin{pmatrix} 1 & -t \\ 0 & 1 \end{pmatrix} \in \underline{\underline{SL}}(2,\mathbb{R})$$

with respect to the $\mathbb{R}$-basis $\{Y_0^*, T_0^*\}$. Thus the G-orbits in $\mathfrak{n}_0^*$ fall into two classes, namely

(I) The single point orbits $\{(\kappa,0) \in \mathfrak{n}_0^* | \kappa \in \mathbb{R}\}$,

and

(II) The line orbits $\{(\mathbb{R}Y_0 + \lambda T_0^*) \subset \mathfrak{n}_0^* | \lambda \in \mathbb{R}, \lambda \neq 0\}$,

as displayed in the following "orbit picture"

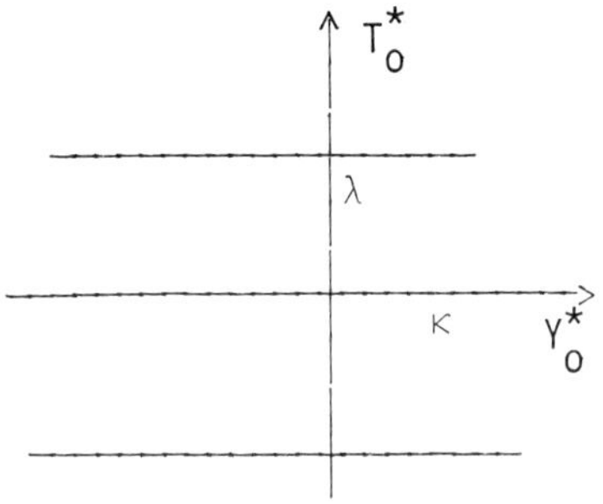

The preceding results can be summarized as follows:

4.19 Theorem. Retain the notations and assumptions of Sections 4.16, 4.17 and 4.18. The closed, normal, abelian subgroup N_0 of the simply connected, nilpotent, real Lie group G is properly embedded into G. If $G_{\hat{y}}$ denotes the stabilizer of $\hat{y} \in \hat{N}_0$ in G and C the centre of G, then either

(I) $G_{\hat{y}} = G$ if and only if $\hat{y}|C = 1$,

or

(II) $G_{\hat{y}} = G_0$ if and only if $\hat{y}|C \neq 1$.

Recall that a real Lie group G is called to be *monomial* if each topologically irreducible, continuous, unitary, linear representation $(U,\mathcal{H})$ of G can be unitarily induced by a one-dimensional, continuous, unitary, linear representation of a suitably chosen closed subgroup H of G (cf. Section 2.17).

The preceding theorem enables us to apply the Mackey machinery to prove the following result.

4.20 Theorem (Dixmier-Kirillov). The simply connected, nilpotent, real Lie groups G are monomial.

Proof. Let C be the centre of G and $(U,\mathcal{H})$ a topologically irreducible, continuous, unitary, linear representation of G. We proceed by induction on the dimension $n \geq 1$ of G. In the case $n = 1$ the statement is obvious. Let $n > 1$ and suppose that the theorem is true for all simply connected, nilpotent, real Lie groups of dimension $< n$.

In case that the dimension of C is ≥ 2 there exists by the reasoning of Section 4.15 a connected, closed, normal Lie subgroup G' of G with dimension ≥ 1 and a topologically irreducible, continuous, unitary, linear representation $(V,\mathcal{H})$ of

$G'' = G/G'$ such that $U = V \circ \pi$ and $\pi: G \to G''$ is the canonical continuous epimorphism. By the inductive hypothesis there exists a connected closed subgroup H'' of G'' such that $(V, \mathcal{H})$ is unitarily induced by a one-dimensional, continuous, unitary, linear representation of H''. By virtue of Lemma 2.14 the given linear representation $(U, \mathcal{H})$ of G is unitarily induced by a one-dimensional, continuous, unitary, linear representation of the subgroup $H = \pi^{-1}(H'')$. Since G' and H'' are connected Lie groups, H is a connected Lie subgroup of G.

It remains to establish the theorem under the additional hypothesis that the centre C of G has dimension 1.

Since G is a unimodular, locally compact, topological group by Theorem 4.14, an application of Theorem 2.20 proves the existence of a character $\hat{x} \in \hat{N}_0$ of N_0 and a topologically irreducible, continuous, unitary, linear representation $(U_0, \mathcal{H}_0)$ of $G_{\hat{x}}$ such that

$$U_0(x) = \hat{x}(x).\mathrm{id}_{\mathcal{H}_0}$$

holds for all $x \in N_0$ and such that $(U, \mathcal{H})$ and $\mathrm{Ind}_{G_{\hat{x}}}^{G}(U_0, \mathcal{H}_0)$ are isomorphic, unitary, linear representations of G. Obviously the identity

$$\mathrm{Ind}_{G_{\hat{x}}}^{G}(U_0, \mathcal{H}_0)(z) = \hat{x}(z).\mathrm{id}_{\mathcal{H}}$$

holds for all elements $z \in C$. If the central character χ_U of $(U, \mathcal{H})$ coincides with the trivial character $\chi_0 = 1$ of C then $G_{\hat{x}} = G$ by Theorem 4.19 and the statement becomes obvious. In the other case we have $\hat{x}|C \neq 1$ and therefore $G_{\hat{x}} = G_0$ by Theorem 4.19. Moreover, G_0 is a simply connected Lie subgroup of G with codimension 1. By the inductive hypothesis there exists a connected closed subgroup H of G_0 and a one-dimensional, continuous, unitary, linear representation of H which induces unitarily $(U_0, \mathcal{H}_0)$. Inducing unitarily $(U, \mathcal{H})$ in stages (Theorem 2.14) proves the result. —

Although the continuous, unitary characters of a simply connected, nilpotent, real Lie group G are, of course, not sufficient to determine its unitary dual $\hat{G}$, the continuous, unitary characters of suitably chosen closed subgroups H of G determine $\hat{G}$ by the unitary inducing procedure. The Kirillov coadjoint orbit picture as exposed in Section 6 infra will show how to choose the subgroups H of G "large enough".

It can be established that, more generally, the simply connected *exponential* real Lie groups are *monomial*. For our purposes, however, the result of

Theorem 4.20 will be sufficient. From its proof we retain the following result.

Corollary. Suppose that the centre C of the simply connected, nilpotent, real Lie group G has dimension 1 and that the central character χ_U of the topologically irreducible, continuous, unitary, linear representation $(U,\mathcal{H})$ of G is different from the trivial character χ_0 of C. Then there exists a simply connected Lie subgroup G_0 of G with codimension 1 and a monomial, topologically irreducible, continuous, unitary, linear representation $(U_0,\mathcal{H}_0)$ of G_0 such that $(U,\mathcal{H})$ and $\mathrm{Ind}_{G_0}^{G}(U_0,\mathcal{H}_0)$ are isomorphic.

In the case of the real Heisenberg nilpotent Lie group the preceding result will allow to determine completely its unitary dual (cf. Theorem 5.6 infra).

4.21 The next step is to reformulate Theorem 4.20 at the Lie algebra level. Let $(U_0,\mathcal{H}_0)$ be a topologically irreducible, continuous, unitary, linear representation of the simply connected, nilpotent, real Lie group G in the complex Hilbert space $\mathcal{H}_0$ and H a suitable connected, closed subgroup of G with a continuous unitary character χ such that

$$(U_0,\mathcal{H}_0) = \mathrm{Ind}_H^G(\chi.\mathrm{id}_{\mathbb{C}},\mathbb{C})$$

holds. Let $\mathfrak{g} = \mathrm{Lie}(G)$ and denote by $\mathfrak{h}$ the Lie subalgebra of the real Lie algebra $\mathfrak{g}$ such that $\mathfrak{h} = \mathrm{Lie}(H)$. Then the differential of χ is a $\mathbb{R}$-linear form $\ell_0 \in \mathfrak{h}^*$ such that

$$\chi(\exp_H X) = e^{2\pi i\langle X,\ell_0\rangle}$$

and

$$\langle [X,Y],\ell_0\rangle = 0$$

holds for all elements $X \in \mathfrak{h}$, $Y \in \mathfrak{h}$. Let $\ell \in \mathfrak{g}^*$ be any $\mathbb{R}$-linear form which extends ℓ_0 to the whole Lie algebra $\mathfrak{g}$. Then $\mathfrak{h}$ forms a totally *isotropic* vector subspace of $\mathfrak{g}$ relative to the *alternating* $\mathbb{R}$-*bilinear form*

$$B_\ell : \mathfrak{g} \times \mathfrak{g} \ni (X,Y) \rightsquigarrow \langle [X,Y],\ell\rangle \in \mathbb{R}$$

associated with ℓ *on* $\mathfrak{g}$. In this case we say that the Lie subalgebra $\mathfrak{h}$ of $\mathfrak{g}$

is *subordinate* to ℓ and we introduce the notation

$$\chi_{\ell,\mathfrak{h}} := \chi,$$

and the *monomial representation*

$$(U_{\ell,\mathfrak{h}}, \mathcal{H}) := \mathrm{Ind}_H^G(\chi_{\ell,\mathfrak{h}} \cdot \mathrm{id}_{\mathbb{C}}, \mathbb{C})$$

of G. Then we obtain

4.22 Theorem. Let G be a simply connected, nilpotent, real Lie group with Lie algebra $\mathfrak{g}$ and $(U_0, \mathcal{H}_0)$ a topologically irreducible, continuous, unitary, linear representation of G in the complex Hilbert space $\mathcal{H}_0$ - then there exists a linear form $\ell \in \mathfrak{g}^*$ and a Lie subalgebra $\mathfrak{h}$ of $\mathfrak{g}$ subordinate to ℓ such that

$$(U_0, \mathcal{H}_0) = (U_{\ell,\mathfrak{h}}, \mathcal{H})$$

holds.

In Section 6 infra we shall study the monomial representations $(U_{\ell,\mathfrak{h}}, \mathcal{H})$ of G in some more detail. It is the purpose of the next section to illustrate our present results in the case of the real Heisenberg nilpotent group. This nilpotent Lie group will be the most important example for our applications.

References

Godement, R. : Introduction à la théorie des groupes de Lie. Tome I et Tome II. Publications Mathématiques de l'Université Paris VII, Volume 11 et Volume 12. Paris 1982.

Sagle, A.A., Walde, R.E. : Introduction to Lie groups and Lie algebras. Pure and Applied Math., Vol. 51. Academic Press, New York, London 1973.

Serre, J.P. : Lie algebras and Lie groups. Mathematics Lecture Note Series. W.A. Benjamin, Reading, Massachusetts 1965.

Varadarajan, V.S. : Lie groups, Lie algebras and their representations. Graduate Texts in Math., Vol. 102. Springer, Berlin, Heidelberg, New York, Tokyo 1984.

5 The real Heisenberg nilpotent Lie group (part I)

5.1 Let $n \geqq 2$ be an integer and denote by E a vector space of dimension n over the commutative field K. A sequence $\mathcal{F} = (E_j)_{0 \leqq j \leqq n}$ of vector subspaces of E forming an ascending filtration

$$E_0 \hookrightarrow E_1 \hookrightarrow E_2 \hookrightarrow \dots \hookrightarrow E_{n-1} \hookrightarrow E_n$$

such that $E_0 = \{0\}$, $E_n = E$, and $\dim_K E_j = j$ for $0 \leqq j \leqq n$ is called to be a *flag* in E. The subgroup $T(\mathcal{F},K)$ of $\mathrm{Aut}(E) = \underline{\underline{GL}}(E)$ consisting of all K-automorphisms u of E such that each vector subspace $E_j (0 \leqq j \leqq n)$ of E is stable under u is called to be the *automorphism group of* $\mathcal{F}$. Since E_{j-1} is of codimension 1 in E_j over K for $1 \leqq j \leqq n$ we can choose a K-basis of E so that any element $u \in T(\mathcal{F},K)$ admits an upper triangular matrix

$$X = \begin{pmatrix} x_{11} & x_{12} & \cdots\cdots & x_{1n} \\ 0 & x_{22} & \cdots\cdots & x_{2n} \\ \cdot & \cdot & & \cdot \\ \cdots & \cdots & \cdots & \cdots \\ \cdot & \cdot & \cdot & \cdot \\ \cdot & \cdot & & \cdot \\ \cdot & \cdot & \cdot & \cdot \\ 0 & 0 & \cdots\cdots & x_{nn} \end{pmatrix}$$

with entries $x_{jk} \in K$ $(1 \leqq j, k \leqq n)$, $x_{jk} = 0$ if $k < j$, and $\det(u) = \det(X) = \prod_{1 \leqq j \leqq n} x_{jj} \neq 0$. We shall denote by $T(n,K)$ the subgroup of $\underline{\underline{GL}}(n,K)$ formed by the upper triangular matrices X of this kind.

Let us consider the case $K = \mathbb{R}$ in some more detail. Since $T(n,\mathbb{R})$ is a closed subgroup of $\underline{\underline{GL}}(n,\mathbb{R})$, we conclude that it forms a real Lie group with right modular function

$$\Delta_{T(n,\mathbb{R})}(X) = \left| \prod_{1 \leqq j \leqq n} x_{jj}^{2j-n-1} \right|.$$

It follows that for an arbitrary flag $\mathcal{F}$ in the n-dimensional real vector space E the automorphism group $T(\mathcal{F},\mathbb{R})$ of $\mathcal{F}$ is a non-unimodular real Lie group.

5.2 Lemma. Let $\mathcal{F}$ denote a flag in the n-dimensional real vector space E - then the real Lie group $T(\mathcal{F},\mathbb{R})$ is solvable.

Proof. Extend the sequence $\mathcal{F} = (E_j)_{0 \leqq j \leqq n}$ to negative indices by setting $E_j = \{0\}$ for $j < 0$ and define the sequence $(\mathfrak{n}_p)_{p \geqq 1}$ of $\mathbb{R}$-subalgebras of the real endomorphism algebra End(E) according to

$$\mathfrak{n}_p = \{u \in \mathrm{End}(E) \,|\, u(E_j) \subseteqq E_{j-p} \textit{ for } 0 \leqq j \leqq n\}.$$

There is a basis of E such that the matrices X_p of the endomorphisms u of E in $\mathfrak{n}_p (p \geqq 1)$ can be put simultaneously into the form $(x_{jk})_{\substack{1 \leqq j \leqq n \\ 1 \leqq k \leqq n}}$ with real entries such that $x_{jk} = 0$ for $k - j < p$. Thus the real matrices X_p are *strictly superdiagonal of order* $p \geqq 1$, that is, have zeros on and below the p-th upper diagonal. Obviously $\mathfrak{n}_n = \{0\}$. Define the sets $(G_p)_{p \geqq 1}$ of endomorphisms of E via the prescription

$$G_p = \mathrm{id}_E + \mathfrak{n}_p \qquad (p \geqq 1).$$

If we set $G_0 = T(\mathcal{F},\mathbb{R})$ then

$$G_0 \hookleftarrow G_1 \hookleftarrow G_2 \hookleftarrow \ldots \hookleftarrow G_{n-1} \hookleftarrow G_n$$

forms a descending filtration of closed normal subgroups of $T(\mathcal{F},\mathbb{R})$ such that $G_n = \{\mathrm{id}_E\}$ and the quotients G_p/G_{p+1} are *abelian* groups for $0 \leqq p \leqq n-1$. It follows $G_p \supseteqq \mathfrak{D}^p T(\mathcal{F},\mathbb{R})$ for $0 \leqq p \leqq n$ and therefore $\mathfrak{D}^n T(\mathcal{F},\mathbb{R}) = \{\mathrm{id}_E\}$. —

Obviously the inclusions

$$\mathfrak{n}_1 \circ \mathfrak{n}_p \subseteqq \mathfrak{n}_{p+1}, \qquad \mathfrak{n}_p \circ \mathfrak{n}_1 \subseteqq \mathfrak{n}_{p+1}$$

hold for $p \geqq 1$. It follows

$$[\mathfrak{n}_1, \mathfrak{n}_p] \subseteqq \mathfrak{n}_{p+1} \qquad (p \geqq 1).$$

In view of Theorem 4.8, the descending filtration of ideals of the real Lie algebra $\mathfrak{n}_1$ given by

$$\mathfrak{n}_1 \hookleftarrow \mathfrak{n}_2 \hookleftarrow \dots \hookleftarrow \mathfrak{n}_{n-1} \hookleftarrow \mathfrak{n}_n$$

shows that

$$\mathfrak{n}(n,\mathbb{R}) := \mathfrak{n}_1$$

is a *nilpotent* Lie algebra over $\mathbb{R}$. It consists of those endomorphisms u of E whose matrix X_1 with respect to a suitable basis of E is *strictly superdiagonal*, that is, has zeros on and below the main diagonal. Define

$$T_1(\mathcal{F},\mathbb{R}) := G_1.$$

Then $T_1(\mathcal{F},\mathbb{R})$ forms a $\frac{1}{2}n(n-1)$-dimensional real Lie group. Its elements are called to be *unipotent automorphisms* of E. There exists a $\mathbb{R}$-basis of E such that the matrices of the unipotent automorphisms of E can be put simultaneously into upper triangular form with 1's on the main diagonal. These *unipotent* matrices with real entries form the closed subgroup $T_1(n,\mathbb{R})$ of $\underline{\underline{GL}}(n,\mathbb{R})$. It follows that $T_1(n,\mathbb{R})$ is a unimodular real Lie group. We have

$$\mathfrak{n}(n,\mathbb{R}) = \mathrm{Lie}(T_1(\mathcal{F},\mathbb{R}))$$

since $\exp_{\underline{\underline{GL}}(n,\mathbb{R})}(tX_1) \in T_1(n,\mathbb{R})$ for all $t \in \mathbb{R}$ and all strictly superdiagonal real matrices X_1.

5.3 Lemma. Let $\mathcal{F}$ denote a flag in the n-dimensional real vector space E - then the real Lie group $T_1(\mathcal{F},\mathbb{R})$ is nilpotent.

Proof. In view of Theorem 4.4 it will be sufficient to establish that

$$[T_1(\mathcal{F},\mathbb{R}),G_p] \subseteqq G_{p+1}$$

holds for $2 \leqq p \leqq n-1$. Let $u \in \mathfrak{n}_1$ and $v \in \mathfrak{n}_p$ be arbitrary elements. Then $(\mathrm{id}_E-u) \in T_1(\mathcal{F},\mathbb{R})$, $(\mathrm{id}_E-v) \in G_p$ are generic elements, and we have

$$\begin{aligned}[(\mathrm{id}_E-u),(\mathrm{id}_E-v)] &= (\mathrm{id}_E-u) \circ (\mathrm{id}_E-v) \circ (\mathrm{id}_E-u)^{-1} \circ (\mathrm{id}_E-v)^{-1} \\ &= (\mathrm{id}_E-u) \circ [(\mathrm{id}_E-u)^{-1} - v \circ (\mathrm{id}_E-u)^{-1}] \circ (\mathrm{id}_E-v)^{-1} \\ &= [\mathrm{id}_E-(\mathrm{id}_E-u) \circ v \circ (\mathrm{id}_E-u)^{-1}] \circ (\mathrm{id}_E-v)^{-1}.\end{aligned}$$

Introduce the endomorphism

$$w = u \cdot v - v \circ \sum_{1 \leqq \nu \leqq n-1} u^{\nu} + u \circ v \circ \sum_{1 \leqq \nu \leqq n-1} u^{\nu}$$

of E. Note that $w \in \mathfrak{n}_{p+1}$ and

$$[(id_E - u), (id_E - v)] = (id_E - v + w) \circ (id_E - v)^{-1}$$

$$= id_E + w \circ (id_E - v)^{-1}$$

$$= id_E + w + \sum_{1 \leqq \nu \leqq n-1} wv^{\nu}.$$

Since the endomorphism $w + \sum_{1 \leqq \nu \leqq n-1} wv^{\nu}$ of E belongs to $\mathfrak{n}_{p+1}$, the proof is complete. —

More general, let a be a real number and $T_a(\mathcal{F},\mathbb{R})$ the subgroup of $T(\mathcal{F},\mathbb{R})$ formed by all automorphisms u of the flag $\mathcal{F} = (E_j)_{0 \leqq j \leqq n}$ in E such that u induces on each one-dimensional quotient vector space E_j/E_{j-1} $(1 \leqq j \leqq n)$ the linear homothetic transformation of ratio a. Then we have the

Corollary. For any real number $a \neq 0$ the real Lie group $T_a(\mathcal{F},\mathbb{R})$ is nilpotent.

5.4 By virtue of the preceding lemma the closed subgroup $\tilde{A}(\mathbb{R}^n)$ of the real Lie group $T_1(n+2,\mathbb{R})$ which is formed by all unipotent real matrices of the form

$$\begin{pmatrix} 1 & x_1 & \cdots & x_n & z \\ 0 & 1 & \cdots & 0 & y_1 \\ \vdots & \vdots & & \vdots & \vdots \\ 0 & 0 & \cdots & 1 & y_n \\ 0 & 0 & \cdots & 0 & 1 \end{pmatrix}$$

is a (2n+1)-dimensional real Lie group. Choose the canonical basis of $\mathbb{R}^n$ and denote by $x \in \mathbb{R}^n$ the vector in the first row with coordinates $(x_j)_{1 \leqq j \leqq n}$ and by $y \in \mathbb{R}^n$ the vector in the last column with coordinates $(y_k)_{1 \leqq k \leqq n}$. Then each element of $\tilde{A}(\mathbb{R}^n)$ admits the form

$$(x,y,z) \in \mathbb{R}^n \times \mathbb{R}^n \times \mathbb{R}.$$

This notation makes it apparent that the underlying manifold of the real Lie group $\tilde{A}(\mathbb{R}^n)$ is given by the vector space $\mathbb{R}^{2n+1}$. The group law of $\tilde{A}(\mathbb{R}^n)$ reads as follows:

$$(x_1,y_1,z_1).(x_2,y_2,z_2) = (x_1+x_2,y_1+y_2,z_1+z_2 + \langle x_1|y_2\rangle),$$

where $(x,y) \rightsquigarrow \langle x|y\rangle = \sum_{1 \leq j \leq n} x_j y_j$ denotes the standard scalar product of the vector space $\mathbb{R}^n$. In the following we shall identify the vector space $\mathbb{R}^n$ with its dual by means of the standard scalar product $\langle\cdot|\cdot\rangle$ and we will consider the column vectors $y \in \mathbb{R}^n$ as linear forms acting on the row vectors $x \in \mathbb{R}^n$. The neutral element $1_{\tilde{A}(\mathbb{R}^n)}$ of $\tilde{A}(\mathbb{R}^n)$ takes the form $(0,0,0)$ and the inverse of an element $(x,y,z) \in \tilde{A}(\mathbb{R}^n)$ is given by

$$(x,y,z)^{-1} = (-x,-y,-z + \langle x|y\rangle).$$

The matrix group $\tilde{A}(\mathbb{R}^n)$ just defined is called the $(2n+1)$-*dimensional real Heisenberg group* in its *dual pairing presentation*. In view of Lemma 5.3 combined with Corollary 1 of Theorem 4.4 supra, $\tilde{A}(\mathbb{R}^n)$ is a simply connected, nilpotent, real Lie group. Its *centre* $\tilde{C}$ is given by

$$\tilde{C} = \{(0,0,z)|z \in \mathbb{R}\}$$

and therefore isomorphic to the additive, locally compact, topological group $\mathbb{R}$. The commutator of any two elements $(x_1,y_1,z_1),(x_2,y_2,z_2)$ of $\tilde{A}(\mathbb{R}^n)$ is given by

$$[(x_1,y_1,z_1),(x_2,y_2,z_2)] = (0,0,B(\begin{bmatrix} x_1 \\ y_1 \end{bmatrix}, \begin{bmatrix} x_2 \\ y_2 \end{bmatrix}))$$

where B denotes the non-degenerate, skew symmetric, $\mathbb{R}$-bilinear form

$$B : (\begin{bmatrix} x_1 \\ y_1 \end{bmatrix}, \begin{bmatrix} x_2 \\ y_2 \end{bmatrix}) \rightsquigarrow \langle x_1|y_2\rangle - \langle y_1|x_2\rangle$$

on $\mathbb{R}^n \oplus \mathbb{R}^n$. We shall call B the *standard symplectic form* on $\mathbb{R}^n \oplus \mathbb{R}^n$ and the pair $(\mathbb{R}^n \oplus \mathbb{R}^n;B)$ the $2n$-*dimensional standard real symplectic vector space* embedded into $\tilde{A}(\mathbb{R}^n)$. It follows

$$\begin{cases} \mathfrak{C}^0\tilde{A}(\mathbb{R}^n) = \mathfrak{C}_2\tilde{A}(\mathbb{R}^n) = \mathfrak{D}^0\tilde{A}(\mathbb{R}^n) = \tilde{A}(\mathbb{R}^n), \\ \mathfrak{C}^1\tilde{A}(\mathbb{R}^n) = \mathfrak{C}_1\tilde{A}(\mathbb{R}^n) = \mathfrak{D}^1\tilde{A}(\mathbb{R}^n) = \tilde{C}, \\ \mathfrak{C}^2\tilde{A}(\mathbb{R}^n) = \mathfrak{C}_0\tilde{A}(\mathbb{R}^n) = \mathfrak{D}^2\tilde{A}(\mathbb{R}^n) = \{1_{\tilde{A}(\mathbb{R}^n)}\}, \end{cases}$$

so that the descending central series and the derived series of $\tilde{A}(\mathbb{R}^n)$ are given by the filtration

$$\tilde{A}(\mathbb{R}^n) \hookleftarrow \tilde{C} \hookleftarrow \{1_{\tilde{A}(\mathbb{R}^n)}\}.$$

Of course, the ascending central series of $\tilde{A}(\mathbb{R}^n)$ takes the form

$$\{1_{\tilde{A}(\mathbb{R}^n)}\} \hookrightarrow \tilde{C} \hookrightarrow \tilde{A}(\mathbb{R}^n).$$

We conclude that $\tilde{A}(\mathbb{R}^n)$ is a *(2n+1)-dimensional, simply connected, two-step nilpotent, real Lie group with one-dimensional centre* $\tilde{C}$. The converse statement that every (2n+1)-dimensional, simply connected, two-step nilpotent, real Lie group having one-dimensional centre is isomorphic to $\tilde{A}(\mathbb{R}^n)$ is also true. Thus $\tilde{A}(\mathbb{R}^n)$ is the simplest possible non-abelian, nilpotent, real Lie group. In particular, $\tilde{A}(\mathbb{R}^n)$ is a *unimodular*, locally compact, topological group by Theorem 4.14 and a *monomial* Lie group by Theorem 4.20.

The Lie algebra $\mathfrak{n}$ of $\tilde{A}(\mathbb{R}^n)$ is the Lie subalgebra of the real nilpotent Lie algebra $\mathfrak{n}(n,\mathbb{R})$ formed by the real nilpotent matrices

$$\begin{pmatrix} 0 & a_1 & \cdots & a_n & c \\ 0 & 0 & \cdots & 0 & b_1 \\ \vdots & & \ddots & & \vdots \\ 0 & 0 & \cdots & 0 & b_n \\ 0 & 0 & \cdots & 0 & 0 \end{pmatrix}$$

We shall call $\mathfrak{n}$ the *(2n+1)-dimensional real Heisenberg nilpotent Lie algebra*. Its centre $\mathfrak{c}$ is formed by the one-dimensional Lie subalgebra of $\mathfrak{n}$ of those matrices which satisfy

$$a_1 = \dots\dots = a_n = 0 = b_1 = \dots\dots = b_n.$$

Denote by $\{X_j, Y_j, T\}$ the canonical basis of the $(2n+1)$-dimensional real vector space $\mathfrak{n}$ such that $\mathfrak{c} = \mathbb{R}T$. Then we have the decomposition

$$\mathfrak{n} = \mathbb{R}^n \oplus \mathbb{R}^n \oplus \mathfrak{c}.$$

Using Kronecker's delta we obtain in $\mathfrak{n}$ *the Heisenberg canonical commutation relations* of quantum mechanics

$$[X_j, Y_k] = \delta_{jk} T, [X_j, X_k] = [Y_j, Y_k] = 0 \qquad (1 \leq j, k \leq n)$$

which explain the name of $\mathfrak{n}$.

Since $\tilde{A}(\mathbb{R}^n)$ is an *exponential* Lie group by Theorem 4.12 supra, the exponential mapping

$$\exp_{\tilde{A}(\mathbb{R}^n)} : \mathfrak{n} \longrightarrow \tilde{A}(\mathbb{R}^n)$$

is a global diffeomorphism which carries the centre $\mathfrak{c}$ of $\mathfrak{n}$ onto the centre $\tilde{C}$ of $\tilde{A}(\mathbb{R}^n)$ and Lebesgue measure $\mu^{\otimes(2n+1)}$ of the $(2n+1)$-dimensional real vector space $\mathfrak{n}$ to Haar measure of $\tilde{A}(\mathbb{R}^n)$.

Apart from the simply connected real Heisenberg nilpotent Lie group $\tilde{A}(\mathbb{R}^n)$ consider the connected Lie group $A(\mathbb{R}^n)$ with underlying manifold $\mathbb{R}^n \times \mathbb{R}^n \times \mathbb{T}$ and group law

$$(x_1, y_1, \zeta_1).(x_2, y_2, \zeta_2) = (x_1 + x_2, y_1 + y_2, \zeta_1 \zeta_2 \, e^{2\pi i \langle x_1 | y_2 \rangle}).$$

We shall call $A(\mathbb{R}^n)$ the $(2n+1)$-dimensional *reduced* Heisenberg nilpotent Lie group in its *dual pairing presentation.* Its centre C is given by

$$C = \{(0,0,\zeta) | \zeta \in \mathbb{T}\}$$

and therefore isomorphic to the one-dimensional compact torus group $\mathbb{T}$. We may consider $A(\mathbb{R}^n)$ as a *central extension* of the additive group $\mathbb{R}^n \oplus \mathbb{R}^n$ by $\mathbb{T}$. Clearly, $\tilde{A}(\mathbb{R}^n)$ is the *universal covering group* of $A(\mathbb{R}^n)$ and both Lie groups have the same Lie algebra $\mathfrak{n}$.

5.5 In order to simplify the notation let us assume $n=1$. For our applications the three-dimensional real Heisenberg nilpotent Lie group $\tilde{A}(\mathbb{R})$ will play the most important rôle. Recall that $\tilde{A}(\mathbb{R})$ may be realized by the three-dimensional Lie group consisting of all unipotent matrices with real entries

$$\begin{pmatrix} 1 & x & z \\ 0 & 1 & y \\ 0 & 0 & 1 \end{pmatrix} = (x,y,z) \in \mathbb{R}^3.$$

The standard symplectic form $B \in \Lambda^2(\mathbb{R} \oplus \mathbb{R})^*$ on the real plane $\mathbb{R} \oplus \mathbb{R}$ is given by

$$B : \left(\begin{bmatrix} x_1 \\ y_1 \end{bmatrix}, \begin{bmatrix} x_2 \\ y_2 \end{bmatrix}\right) \to \det \begin{pmatrix} x_1 & x_2 \\ y_1 & y_2 \end{pmatrix}$$

so that in the dual pairing presentation of $\tilde{A}(\mathbb{R})$ the commutator of the elements (x_1,y_1,z_1), (x_2,y_2,z_2) of $\tilde{A}(\mathbb{R})$ takes the form

$$[(x_1,y_1,z_1),(x_2,y_2,z_2)] = (0,0, \det \begin{pmatrix} x_1 & x_2 \\ y_1 & y_2 \end{pmatrix}) \in \tilde{C}.$$

The three-dimensional Lie algebra $\mathfrak{n}$ of $\tilde{A}(\mathbb{R})$ is formed by the nilpotent matrices

$$\begin{pmatrix} 0 & a & c \\ 0 & 0 & b \\ 0 & 0 & 0 \end{pmatrix}$$

with real entries a,b,c. The canonical basis $\{X,Y,T\}$ of $\mathfrak{n}$ with $\mathfrak{c} = \mathbb{R}.T$ will be formed by the matrices

$$X = \begin{pmatrix} 0 & 1 & 0 \\ 0 & 0 & 0 \\ 0 & 0 & 0 \end{pmatrix} \quad Y = \begin{pmatrix} 0 & 0 & 0 \\ 0 & 0 & 1 \\ 0 & 0 & 0 \end{pmatrix}, \quad T = \begin{pmatrix} 0 & 0 & 1 \\ 0 & 0 & 0 \\ 0 & 0 & 0 \end{pmatrix},$$

so that $[X,Y] = T$ and all the other brackets vanish. If $\{X^*,Y^*,T^*\}$ denotes the dual basis of $\mathfrak{n}^*$, then each linear form $\ell \in \mathfrak{n}^*$ takes the form

$$\ell = \rho X^* + \kappa Y^* + \lambda T^*$$

with real coordinates $(\rho,\kappa,\lambda) \in \mathbb{R}^3$.

As observed earlier, $\tilde{A}(\mathbb{R})$ is a monomial real Lie group. In order to determine a complete set of representatives of the unitary dual $\tilde{A}(\mathbb{R})\hat{}$ of $\tilde{A}(\mathbb{R})$ by the Mackey machinery, note that $\tilde{A}(\mathbb{R})$ is the semi-direct product of the closed, normal, maximal abelian subgroup N_0 given by

$$N_0 = \{(0,y,z) | y \in \mathbb{R},\ z \in \mathbb{R}\}$$

with the closed subgroup K_0 of $\tilde{A}(\mathbb{R})$ where

$$K_0 = \{(x,0,0) | x \in \mathbb{R}\}.$$

The action of K_0 on N_0 is given by

$$(x,0,0)(0,y,z)(-x,0,0) = (0,y,z + xy)$$

and the corresponding Lie subalgebras $\mathfrak{n}_0 = \mathrm{Lie}(N_0)$, $\mathfrak{k}_0 = \mathrm{Lie}(K_0)$ of $\mathfrak{n}$ take the form

$$\mathfrak{n}_0 = \mathbb{R}Y \oplus \mathfrak{c},$$

$$\mathfrak{k}_0 = \mathbb{R}X.$$

The ideal $\mathfrak{n}_0$ of $\mathfrak{n}$ forms the centralizer of Y in $\mathfrak{n}$ and

$$\mathfrak{n} = \mathfrak{k}_0 \oplus \mathfrak{n}_0.$$

The mapping $\pi:\tilde{A}(\mathbb{R}) \ni (x,y,z) \rightsquigarrow x \in \mathbb{R}$ defines a continuous epimorphism of the real Heisenberg nilpotent Lie group $\tilde{A}(\mathbb{R})$ onto the additive Lie group $\mathbb{R}$ with kernel N_0. It induces a bicontinuous isomorphism of $\tilde{A}(\mathbb{R})/N_0$ onto $\mathbb{R}$. If we identify the groups $\mathbb{R}$ and K_0 by means of the continuous section $s : x \rightsquigarrow (x,0,0)$ of π, then the canonical epimorphism $\tilde{A}(\mathbb{R}) \rightsquigarrow \tilde{A}(\mathbb{R})/N_0$ identifies with the mapping $\tilde{A}(\mathbb{R}) \ni (x,y,z) \rightsquigarrow (x,0,0) \in K_0$ and s is a continuous section of $\tilde{A}(\mathbb{R})$ fibred by N_0.

For any pair $(\kappa,\lambda) \in \mathbb{R} \times \mathbb{R}$ the function

$$\chi_{(\kappa,\lambda)} : N_0 \ni (0,y,z) \rightsquigarrow e^{2\pi i(\kappa y+\lambda z)} \in \mathbb{T}$$

defines a continuous, unitary character of N_0. The left action of $\tilde{A}(\mathbb{R})$ on $\hat{N}_0$ by inner automorphisms of N_0 is given by (cf. Sections 4.17 and 4.18)

$$(x,y,z).\chi_{(\kappa,\lambda)} = \chi_{(\kappa-x\lambda,\lambda)}.$$

Consistently with 4.18 and the orbit picture displayed there, the $\tilde{A}(\mathbb{R})$-orbits in $\hat{N}_0$ fall into two classes, namely

(I) The orbits $\{\chi_{(\kappa,0)}|\kappa \in \mathbb{R}\}$ which may be identified with the single point sets $\{(\kappa,0)|\kappa \in \mathbb{R}\}$ in $\mathbb{R} \oplus \mathbb{R}$,

and

(II) The orbits $\{\{\chi_{(\mu,\lambda)}|\mu \in \mathbb{R}\}|\lambda \in \mathbb{R},\ \lambda \neq 0\}$ which may be identified with the lines $\mathbb{R} \times \{\lambda\}$, $\lambda \neq 0$, in $\mathbb{R} \oplus \mathbb{R}$.

It follows again that N_0 is *properly embedded* into $\tilde{A}(\mathbb{R})$.

(I) The stabilizer of $\chi_{(\kappa,0)}$ in $\tilde{A}(\mathbb{R})$ is the whole group $\tilde{A}(\mathbb{R})$ for all $\kappa \in \mathbb{R}$. Therefore the elements of $\tilde{A}(\mathbb{R})^\wedge$ associated with the single point orbit $\{(\kappa,0)|\kappa \in \mathbb{R}\}$ are the isomorphy classes of the unitary characters given by

$$U_{(\rho,\kappa)}:\tilde{A}(\mathbb{R}) \ni (x,y,z) \rightsquigarrow e^{2\pi i(\rho x+\kappa y)} \in \mathbb{T} \quad ((\rho,\kappa) \in \mathbb{R} \times \mathbb{R}).$$

The one-dimensional, continuous, unitary, linear representations $(U_{(\rho,\kappa)},\mathbb{C})$ of $\tilde{A}(\mathbb{R})$ where $(\rho,\kappa) \in \mathbb{R} \times \mathbb{R}$ are called to be the *degenerate* linear representations of $\tilde{A}(\mathbb{R})$. Their isomorphy classes are collected in the subset $\tilde{A}(\mathbb{R})^\wedge_{(I)}$ of $\tilde{A}(\mathbb{R})$.

(II) The stablilizer in $\tilde{A}(\mathbb{R})$ for the *generic* element $(0,\lambda)$ with $\lambda \neq 0$ of the line orbit $\mathbb{R} \times \{\lambda\}$ is the subgroup N_0 itself and by Theorem 2.18 there is only one element of $\tilde{A}(\mathbb{R})^\wedge$ associated with $\mathbb{R} \times \{\lambda\}$ namely the isomorphy class of the monomial representation

$$(U_\lambda,\mathcal{H}) = \mathrm{Ind}_{N_0}^{\tilde{A}(\mathbb{R})}(\chi_{(0,\lambda)}\cdot\mathrm{id}_{\mathbb{C}},\mathbb{C}) \quad (\lambda \neq 0)$$

of $\tilde{A}(\mathbb{R})$. Since every function $\phi \in \mathcal{K}(\tilde{A}(\mathbb{R})/N_0,\mathbb{C})$ is determined uniquely by its restriction to the closed subgroup K_0 of $\tilde{A}(\mathbb{R})$ and since conversely every function $\psi \in \mathcal{K}(K_0)$ gives rise to the function

$$\phi:\ (x,y,z) \rightsquigarrow e^{2\pi i(\mu y+\lambda(z-xy))}\psi((x,0,0))$$

which belongs to $\mathcal{K}(\tilde{A}(\mathbb{R})/N_0,\mathbb{C})$ for $(\mu,\lambda) \in \mathbb{R} \times \mathbb{R}$, $\lambda \neq 0$, we may identify the complex vector spaces $\mathcal{K}(\tilde{A}(\mathbb{R})/N_0,\mathbb{C})$ and $\mathcal{K}(K_0)$. It follows that the representation space $\mathcal{H}$ of U_λ is the complex Hilbert space $L^2(\mathbb{R})$ for all $\lambda \neq 0$. If $\ell \in \mathfrak{n}^*$ denotes a linear form on $\mathfrak{n}$ such that $\langle T,\ell\rangle = \lambda \neq 0$ - then $\mathfrak{n}_0$ and $\mathfrak{k}_0$ are Lie subalgebras of $\mathfrak{n}$ subordinate to ℓ and we have in the case $\langle X,\ell\rangle = 0$ the identity

$$(U_\lambda,\mathcal{H}) = (U_{\ell,\mathfrak{n}_0},L^2(\mathbb{R})) \quad (\lambda \neq 0).$$

Moreover, we obtain the action

$$U_\lambda(x,y,z)\psi(t) = e^{2\pi i\lambda(z-yt)}\psi(t-x) \qquad (t \in \mathbb{R})$$

of $\tilde{A}(\mathbb{R})$ on the functions $\psi \in \mathbb{K}(\mathbb{R})$. In order to prove that the *non-degenerate* linear representations $(U_\lambda, L^2(\mathbb{R}))$ of $\tilde{A}(\mathbb{R})$ are topologically irreducible, continuous, unitary, linear representations of $\tilde{A}(\mathbb{R})$ for all values of the real parameter $\lambda \neq 0$ we shall establish the following result (cf. Corollary 2 of Theorem 1.6 supra) by using elementary distribution theory on the real line $\mathbb{R}$.

5.6 Theorem. For $\lambda \in \mathbb{R}$, $\lambda \neq 0$, the centralizer in $\underline{\underline{U}}(L^2(\mathbb{R}))$ of the image group $U_\lambda(\tilde{A}(\mathbb{R}))$ of $\tilde{A}(\mathbb{R})$ under U_λ satisfies

$$\mathfrak{R}_{\tilde{A}(\mathbb{R})}(U_\lambda) \cap \underline{\underline{U}}(L^2(\mathbb{R})) = \mathbb{T}.\mathrm{id}_{L^2(\mathbb{R})}.$$

Proof. Let $T \in \mathfrak{R}_{\tilde{A}(\mathbb{R})}(U_\lambda) \cap \underline{\underline{U}}(L^2(\mathbb{R}))$ denote a unitary automorphism of $(U_\lambda, L^2(\mathbb{R}))$ where $\lambda \neq 0$ is a fixed real parameter value. If ϕ and ψ denote functions belonging to $\mathbb{K}(\mathbb{R})$ then $(T\phi) * \psi$ and $T(\phi*\psi)$ both belong to $L^2(\mathbb{R})$ and an application of Fubini's theorem yields for any $\omega \in L^2(\mathbb{R})$ the identities

$$\begin{aligned}
\langle (T\phi) * \psi | \omega\rangle &= \int_{\mathbb{R}} (T\phi) * \psi(t)\bar{\omega}(t)d\mu(t) \\
&= \int_{\mathbb{R}} [\int_{\mathbb{R}} (T\phi)(t-x)\psi(x)d\mu(x)]\bar{\omega}(t)d\mu(t) \\
&= \int_{\mathbb{R}} \psi(x) [\int_{\mathbb{R}} (U_\lambda(x,0,0) \circ T\phi)(t)\bar{\omega}(t)d\mu(t)]d\mu(x) \\
&= \int_{\mathbb{R}} \psi(x) [\int_{\mathbb{R}} (T \circ U_\lambda(x,0,0)\phi)(t)\bar{\omega}(t)d\mu(t)]d\mu(x) \\
&= \int_{\mathbb{R}} \psi(x) [\int_{\mathbb{R}} U_\lambda(x,0,0)\phi(t)T^*\bar{\omega}(t)d\mu(t)]d\mu(x) \\
&= \int_{\mathbb{R}} [\int_{\mathbb{R}} U_\lambda(x,0,0)\phi(t)\psi(x)d\mu(x)]T^*\bar{\omega}(t)d\mu(t) \\
&= \int_{\mathbb{R}} \phi*\psi(t)T^*\bar{\omega}(t)d\mu(t) \\
&= \int_{\mathbb{R}} T(\phi*\psi)(t)\bar{\omega}(t)d\mu(t) \\
&= \langle T(\phi*\psi)|\omega\rangle.
\end{aligned}$$

Since these identities hold for all elements $\omega \in L^2(\mathbb{R})$, an easy argument using the everywhere denseness of $\mathbb{K}(\mathbb{R})$ in $L^1(\mathbb{R}) \cap L^2(\mathbb{R})$ shows that

$$T(\phi * \psi) = (T\phi) * \psi = \phi*(T\psi)$$

holds for ϕ and ψ in $L^1(\mathbb{R}) \cap L^2(\mathbb{R})$. An application of the Fourier transform $\mathcal{F}_{\mathbb{R}}$ yields

$$\mathcal{F}_{\mathbb{R}}(T\phi).\mathcal{F}_{\mathbb{R}}\psi = \mathcal{F}_{\mathbb{R}}\phi.\mathcal{F}_{\mathbb{R}}(T\psi).$$

For any $t \in \mathbb{R}$ there exists an element $\phi \in L^1(\mathbb{R}) \cap L^2(\mathbb{R})$ such that $\mathcal{F}_{\mathbb{R}}\phi(t) \neq 0$. Define

$$f(t) = \frac{\mathcal{F}_{\mathbb{R}}(T\phi)(t)}{\mathcal{F}_{\mathbb{R}}\phi(t)} .$$

The identity just obtained above shows that this definition is independent of the choice of $\phi \in L^1(\mathbb{R}) \cap L^2(\mathbb{R})$, and, moreover, that

$$\mathcal{F}_{\mathbb{R}}(T\phi) = f.\mathcal{F}_{\mathbb{R}}\phi$$

for each $\phi \in L^1(\mathbb{R}) \cap L^2(\mathbb{R})$. Clearly f is unique. Furthermore f belongs locally to $L^2(\mathbb{R})$. From this result we can easily deduce that the equivalence class of Lebesgue measurable functions on $\mathbb{R}$ defined by f satisfies

$$||f||_{\infty} = 1.$$

An argument using the denseness of $L^1(\mathbb{R}) \cap L^2(\mathbb{R})$ in $L^2(\mathbb{R})$, the Plancherel theorem, and the boundedness of f shows that

$$\mathcal{F}_{\mathbb{R}}(T\Phi) = f.\mathcal{F}_{\mathbb{R}}\Phi$$

for each $\Phi \in L^2(\mathbb{R})$. There exists a unique tempered distribution $S \in \mathcal{S}'(\mathbb{R})$ such that $\mathcal{F}_{\mathbb{R}}S = f$. It follows

$$T\Phi = S * \phi$$

for each continuous function $\Phi:\mathbb{R} \to \mathbb{C}$ rapidly decreasing at infinity. In view of $T \circ U_{\lambda}(0,y,0) = U_{\lambda}(0,y,0) \circ T$ for all $y \in \mathbb{R}$, the assumption $\lambda \neq 0$ implies

$$T(\chi\phi) = \chi(T\phi) = S * (\chi\phi) = \chi(S * \phi) = (\chi S) * (\chi\phi)$$

for all continuous unitary characters $\chi \in \hat{\mathbb{R}}$ of the additive locally compact group $\mathbb{R}$. It follows

$$(1-\chi)S = 0 \qquad (\chi \in \hat{\mathbb{R}})$$

and therefore S has its support at the origin of $\mathbb{R}$. In view of $\|\mathcal{F}_{\mathbb{R}}S\|_\infty = \|f\|_\infty = 1$ we conclude that

$$S = \zeta.\varepsilon_0$$

for a unique number $\zeta \in \mathbb{T}$. Thus $T = \zeta.\mathrm{id}_{L^2(\mathbb{R})}$. —

If $\tilde{A}(\mathbb{R})^\wedge_{(II)}$ denotes the set of isomorphy classes of the family $\{(U_\lambda, L^2(\mathbb{R})) \mid \lambda \in \mathbb{R}, \lambda \neq 0\}$ of non-degenerate, topologically irreducible, continuous, unitary, linear representations of $\tilde{A}(\mathbb{R})$ then we have

5.7 <u>Theorem</u>. The unitary dual of the real Heisenberg nilpotent Lie group $\tilde{A}(\mathbb{R})$ admits the decomposition into disjoint subsets

$$\tilde{A}(\mathbb{R})^\wedge = \tilde{A}(\mathbb{R})^\wedge_{(I)} \cup \tilde{A}(\mathbb{R})^\wedge_{(II)}$$

where the subset $\tilde{A}(\mathbb{R})^\wedge_{(I)}$ can be bijectively mapped onto the dual $(\mathbb{R} \oplus \mathbb{R})^\wedge$ and the subset $\tilde{A}(\mathbb{R})^\wedge_{(II)}$ is in bijective correspondence with the set $\hat{\tilde{C}} - \{0\}$ of non-trivial central characters of $\tilde{A}(\mathbb{R})$.

5.8 Let us agree to retain the above notations. In case when we use the *right* action of $\tilde{A}(\mathbb{R})$ on $\mathbb{K}(N_0 \backslash \tilde{A}(\mathbb{R}), \mathbb{C})$ we obtain for all $\lambda \in \mathbb{R}, \lambda \neq 0$, the realization

$$U_\lambda'(x,y,z)\psi(t) = e^{2\pi i\lambda(z+yt)}\psi(t+x) \qquad (t \in \mathbb{R})$$

of $(U_\lambda, L^2(\mathbb{R}))$ acting on the functions $\psi \in \mathbb{K}(\mathbb{R})$. Needless to say that the continuous, unitary, linear representations $(U_\lambda, L^2(\mathbb{R}))$ and $(U_\lambda', L^2(\mathbb{R}))$ of $\tilde{A}(\mathbb{R})$ are unitarily isomorphic for all real numbers $\lambda \neq 0$. Finally it should be observed that $\tilde{A}(\mathbb{R})$ is the semi-direct product of the closed, normal, maximal abelian subgroup M_0 given by

$$M_0 = \{(x,0,z) \mid x \in \mathbb{R}, z \in \mathbb{R}\}$$

with the closed subgroup L_0 defined by

$$L_0 = \{(0,y,0) \mid y \in \mathbb{R}\}.$$

The action of L_0 on M_0 is given by

$$(0,y,0)(x,0,z)(0,-y,0) = (x,0,z-xy).$$

Obviously the automorphism of $\tilde{A}(\mathbb{R})$ defined by the assignment

$$J : (x,y,z) \rightsquigarrow (y,-x,z-xy)$$

has the inverse

$$J^{-1} : (x,y,z) \rightsquigarrow (-y,x,z-xy).$$

It follows $U'_\lambda = U_\lambda \circ J^2$. Furthermore, J maps N_0 onto M_0, and K_0 onto L_0, and leaves the centre $\tilde{C}$ of $\tilde{A}(\mathbb{R})$ pointwise fixed. The corresponding Lie subalgebras $\mathfrak{m}_0 = \mathrm{Lie}(M_0)$ and $\mathfrak{l}_0 = \mathrm{Lie}(L_0)$ of $\mathfrak{n}$ are given by

$$\mathfrak{m}_0 = \mathbb{R}X \oplus \mathbb{R}T,$$

$$\mathfrak{l}_0 = \mathbb{R}Y.$$

The ideal $\mathfrak{m}_0$ of $\mathfrak{n}$ forms the centralizer of X in $\mathfrak{n}$ and we have

$$\mathfrak{n} = \mathfrak{l}_0 \oplus \mathfrak{m}_0.$$

If $\ell \in \mathfrak{n}^*$ denotes a linear form on $\mathfrak{n}$ such that $\langle T,\ell\rangle = \lambda \neq 0$ - then $\mathfrak{m}_0$ and $\mathfrak{l}_0$ are Lie subalgebras of $\mathfrak{n}$ subordinate to ℓ. Suppose $\langle Y,\ell\rangle = 0$. The monomial representation

$$(V_\lambda,\mathcal{H}) = \mathrm{Ind}_{M_0}^{\tilde{A}(\mathbb{R})}(\chi_{(0,\lambda)}\cdot \mathrm{id}_{\mathbb{C}},\mathbb{C}) \qquad (\lambda \neq 0)$$

$$= (U_{\ell,\mathfrak{m}_0}, L^2(\mathbb{R}))$$

of $\tilde{A}(\mathbb{R})$ is realized by $(V_\lambda, L^2(\mathbb{R}))$ where $\tilde{A}(\mathbb{R})$ acts by $V_\lambda = U_\lambda \circ J$ on the functions $\psi \in \mathcal{K}(\mathbb{R})$ according to the prescription

$$V_\lambda(x,y,z)\psi(s) = e^{2\pi i\lambda(z+xs-xy)}\psi(s-y) \qquad (s \in \mathbb{R})$$

and by $(V'_\lambda, L^2(\mathbb{R}))$ where $\tilde{A}(\mathbb{R})$ acts by $V'_\lambda = U'_\lambda \circ J$ on the functions

$\psi \in \mathbb{K}(\mathbb{R})$ according to the rule

$$V'_\lambda(x,y,z)\psi(s) = e^{2\pi i\lambda(z-xs-xy)}\psi(s+y) \qquad (s \in \mathbb{R}).$$

Again, it is obvious that the continuous, unitary, linear representations $(V_\lambda, L^2(\mathbb{R}))$ and $(V'_\lambda, L^2(\mathbb{R}))$ of $\tilde{A}(\mathbb{R})$ are unitarily isomorphic. A direct calculation shows that the bicontinuous linear bijection

$$\mathcal{F}^\lambda_{\mathbb{R}} : L^2(\mathbb{R}) \to L^2(\mathbb{R}) \qquad (\lambda \neq 0)$$

which acts on the functions $\psi \in \mathbb{K}(\mathbb{R})$ according to the prescription

$$\mathcal{F}^\lambda_{\mathbb{R}}\psi(t) = \int_{\mathbb{R}} \psi(s)\, e^{-2\pi i\lambda ts} d\mu(s) \qquad (t \in \mathbb{R})$$

extends to an isomorphism of $(V_\lambda, L^2(\mathbb{R}))$ onto $(U_\lambda, L^2(\mathbb{R}))$. Note that $\mathcal{F}^1_{\mathbb{R}}$ is the *Fourier transform* $\mathcal{F}_{\mathbb{R}}$. It is well known that $\mathcal{F}_{\mathbb{R}}$ forms a unitary automorphism of the complex Hilbert space $L^2(\mathbb{R})$ having as its inverse the *Fourier cotransform* $\bar{\mathcal{F}}_{\mathbb{R}}$ which acts on the functions $\psi \in \mathbb{K}(\mathbb{R})$ via the prescription

$$\bar{\mathcal{F}}_{\mathbb{R}}\psi(s) = \int_{\mathbb{R}} \psi(t) e^{2\pi ist} d\mu(t) \qquad (s \in \mathbb{R})$$

and extends to a unitary isomorphism of $(U'_1, L^2(\mathbb{R}))$ onto $(V'_1, L^2(\mathbb{R}))$. It follows that for all values of the parameter $\lambda \in \mathbb{R}$, $\lambda \neq 0$, the *non-degenerate*, topologically irreducible, continuous, unitary, linear representations

$$(U_\lambda, L^2(\mathbb{R})), \quad (U'_\lambda, L^2(\mathbb{R})),$$

$$(V_\lambda, L^2(\mathbb{R})), \quad (V'_\lambda, L^2(\mathbb{R}))$$

of $\tilde{A}(\mathbb{R})$ are unitarily isomorphic. Their isomorphy class belongs to the subset $\tilde{A}(\mathbb{R})^\wedge_{(II)}$ of the unitary dual $\tilde{A}(\mathbb{R})^\wedge$ and admits the central character

$$\chi_\lambda := \chi_{(0,\lambda)} : \tilde{C} \ni (0,0,z) \rightsquigarrow \chi_1(0,0,\lambda z) \in \mathbb{T} \qquad (\lambda \neq 0)$$

where

$$\chi_1 := \chi_{(0,1)} : \tilde{C} \ni (0,0,z) \rightsquigarrow e^{2\pi iz} \in \mathbb{T}$$

is the *basic character of* $\tilde{C}$. The corresponding vector subspace of $L^2(\mathbb{R})$ of *smooth* vectors (cf. Section 1.2, Remark 3)

$$L^2(\mathbb{R})^\infty = \mathcal{S}(\mathbb{R})$$

is the *Schwartz space* on the real line $\mathbb{R}$, i.e., the complex vector space $\mathcal{S}(\mathbb{R})$ of all *infinitely differentiable, complex-valued functions on* $\mathbb{R}$ *that are rapidly vanishing at infinity.* In view of the ascending filtration

$$K(\mathbb{R}) \hookrightarrow \mathcal{S}(\mathbb{R}) \hookrightarrow L^2(\mathbb{R})$$

the Schwartz space is an everywhere dense vector subspace of the complex Hilbert space $L^2(\mathbb{R})$.

An extension of the results of this section to the (2n+1)-dimensional, real Heisenberg nilpotent Lie groups $\tilde{A}(\mathbb{R}^n)$, $n > 1$, is straightforward.

5.9 Apart from the dual pairing presentation of the (2n+1)-dimensional real Heisenberg nilpotent Lie group $\tilde{A}(\mathbb{R}^n)$ there exists a "symmetrized" presentation which is based on the standard symplectic form B on the 2n-dimensional real vector space $\mathbb{R}^n \oplus \mathbb{R}^n$. Indeed, let the real vector space $(\mathbb{R}^n \oplus \mathbb{R}^n) \times \mathbb{R}$ be equipped with the multiplication law

$$(x_1,y_1,z_1).(x_2,y_2,z_2) = (x_1+x_2,y_1+y_2,z_1+z_2 + B(\begin{bmatrix} x_1 \\ y_1 \end{bmatrix}, \begin{bmatrix} x_2 \\ y_2 \end{bmatrix})).$$

The fact that the kernel of a skew symmetric, bilinear form B on $\mathbb{R}^n \oplus \mathbb{R}^n$ is the *only* invariant of the form implies that the (2n+1)-dimensional, two-step nilpotent, real Lie group which arises in this way is isomorphic with $\tilde{A}(\mathbb{R}^n)$. It will be called the (2n+1)-*dimensional real Heisenberg nilpotent Lie group* in its *basic presentation* since it is of particular importance in applications. If we replace the dual pairing presentation by the basic presentation of $\tilde{A}(\mathbb{R}^n)$, the non-degenerate, continuous, unitary, linear representations $(U_\lambda, L^2(\mathbb{R}^n))$ and $(U'_\lambda, L^2(\mathbb{R}^n))$ of $\tilde{A}(\mathbb{R}^n)$ with central character χ_λ considered in Sections 5.5 and 5.8 act on the functions $\psi \in \mathcal{S}(\mathbb{R}^n)$ as follows:

$$U_\lambda(x,y,z)\psi(t) = e^{2\pi i\lambda(z-\langle y|t\rangle + \frac{1}{2}\langle x|y\rangle)}\psi(t-x),$$

$$U'_\lambda(x,y,z)\psi(t) = e^{2\pi i\lambda(z+\langle y|t\rangle + \frac{1}{2}\langle x|y\rangle)}\psi(t+x), \qquad (t \in \mathbb{R}^n)$$

where $\lambda \neq 0$ is a real parameter and $(x,y,z) \in \tilde{A}(\mathbb{R}^n)$. In the next section we will see that $(U_\lambda, L^2(\mathbb{R}^n))$ and $(U'_\lambda, L^2(\mathbb{R}^n))$ are unitarily isomorphic, topologically irreducible, continuous, unitary, linear representations of $\tilde{A}(\mathbb{R}^n)$

in its basic presentation for all values of the real parameter $\lambda \neq 0$. The automorphism

$$J : (x,y,z) \rightsquigarrow (y,-x,z)$$

of $\tilde{A}(\mathbb{R}^n)$ which leaves the centre $\tilde{C}$ pointwise fixed and its inverse

$$J^{-1} : (x,y,z) \rightsquigarrow (-y,x,z)$$

give rise to the unitarily isomorphic, topologically irreducible, continuous, unitary, linear representations $(V_\lambda, L^2(\mathbb{R}^n))$ and $(V'_\lambda, L^2(\mathbb{R}^n))$ of $\tilde{A}(\mathbb{R}^n)$ where $V_\lambda = U_\lambda \circ J$ acts on the functions $\psi \in \mathcal{S}(\mathbb{R}^n)$ according to the prescription

$$V_\lambda(x,y,z)\psi(s) = e^{2\pi i\lambda(z+\langle x|s\rangle-\frac{1}{2}\langle x|y\rangle)}\psi(s-y) \qquad (s \in \mathbb{R}^n)$$

and $V' = U'_\lambda \circ J$ acts on the functions $\psi \in \mathcal{S}(\mathbb{R}^n)$ according to the rule

$$V'_\lambda(x,y,z)\psi(s) = e^{2\pi i\lambda(z-\langle x|s\rangle-\frac{1}{2}\langle x|y\rangle)}\psi(s+y) \qquad (s \in \mathbb{R}^n)$$

for $(x,y,z) \in \tilde{A}(\mathbb{R}^n)$. Of course, $U_\lambda = U'_\lambda \circ J^2$. If the real parameter $\lambda \neq 0$ is normalized to $\lambda = 1$, then the Fourier transform $\mathcal{F}_{\mathbb{R}^n} : \mathcal{S}(\mathbb{R}^n) \to \mathcal{S}(\mathbb{R}^n)$ defined by

$$\mathcal{F}_{\mathbb{R}^n}\psi(t) = \int_{\mathbb{R}^n} \psi(s)e^{-2\pi i\langle t|s\rangle}d\mu^{\otimes n}(s) \qquad (t \in \mathbb{R}^n),$$

extends to a unitary isomorphism of $(V_1, L^2(\mathbb{R}^n))$ onto $(U_1, L^2(\mathbb{R}^n))$ and its inverse, the Fourier cotransform $\bar{\mathcal{F}}_{\mathbb{R}^n} : \mathcal{S}(\mathbb{R}^n) \to \mathcal{S}(\mathbb{R}^n)$ defined by

$$\bar{\mathcal{F}}_{\mathbb{R}^n}\psi(s) = \int_{\mathbb{R}^n} \psi(t)e^{2\pi i\langle s|t\rangle}d\mu^{\otimes n}(t) \qquad (s \in \mathbb{R}^n),$$

extends to a unitary isomorphism of $(U'_1, L^2(\mathbb{R}^n))$ onto $(V'_1, L^2(\mathbb{R}^n))$.

References

Auslander, L. : Lecture notes on nil-theta functions. Regional Conference Series in Mathematics, No. 34. American Mathematical Society, Providence, Rhode Island 1977.

Howe, R. : On the role of the Heisenberg group in harmonic analysis. Bull. (New Series) Amer. Math. Soc. 3 (1980), 821-843.

Van der Waerden, B.L. : Sources of quantum mechanics. North-Holland, Amsterdam 1967.

Weyl, H. : Gruppentheorie und Quantenmechanik. Wissenschafliche Buchgesellschaft, Darmstadt 1981.

6 The coadjoint orbit picture

6.1 One of the most powerful geometric ideas in the representation theory of simply connected Lie groups G is the Kirillov orbit method which seeks to relate the topologically irreducible, continuous, unitary, linear representations of G to the set $\mathfrak{g}^*/G$ of orbits of G in the dual vector space $\mathfrak{g}^*$ of the Lie algebra $\mathfrak{g} = \mathrm{Lie}(G)$. In the following G denotes a simply connected, nilpotent, real Lie group. Form the real vector space $\mathfrak{g}^*$ of all $\mathbb{R}$-linear forms ℓ on $\mathfrak{g}$ and the associated real Lie group $\underline{\underline{GL}}(\mathfrak{g}^*)$. Recall from Section 4.17 supra that for all elements $x \in G$ the $\mathbb{R}$-linear mapping

$$\mathrm{CoAd}_G(x) : \mathfrak{g}^* \to \mathfrak{g}^*$$

is the *contragredient* of the coadjoint linear mapping $\mathrm{Ad}_G(x) : \mathfrak{g} \to \mathfrak{g}$, i.e., the *transpose* of the $\mathbb{R}$-linear mapping

$$\mathrm{Ad}_G(x^{-1}) : \mathfrak{g} \to \mathfrak{g}.$$

The Lie group morphism $\mathrm{CoAd}_G : G \to \underline{\underline{GL}}(\mathfrak{g}^*)$ is called to be the *coadjoint linear representation* of G. For an arbitrary linear form $\ell \in \mathfrak{g}^*$ denote by G_ℓ the *stabilizer* of ℓ under the coadjoint linear action of G on $\mathfrak{g}^*$. Thus we have

$$G_\ell = \{x \in G \mid \mathrm{CoAd}_G(x)\ell = \ell\} .$$

Since G_ℓ is a closed subgroup of G it forms a Lie subgroup of G, hence a nilpotent real Lie group. Let $\mathfrak{g}_\ell = \mathrm{Lie}(G_\ell)$ - then $\mathfrak{g}_\ell$ is a Lie subalgebra of $\mathfrak{g}$ and we have for all $X \in \mathfrak{g}_\ell$, $Y \in \mathfrak{g}$, and the parameter $t \in \mathbb{R}$ the identities

$$\begin{aligned} \langle Y,\ell\rangle &= \langle Y, \mathrm{CoAd}_G(\exp_G tX)\ell\rangle \\ &= \langle \mathrm{Ad}_G(\exp_G tX)^{-1} Y, \ell\rangle \\ &= \langle \exp_G(-t.\mathrm{ad}_{\mathfrak{g}} X)Y, \ell\rangle. \end{aligned}$$

Forming the derivative with respect to t at the origin t = 0 yields

$$\langle [X,Y],\ell\rangle = 0$$

for all $X \in \mathfrak{g}_\ell$ and $Y \in \mathfrak{g}$. In order to interpret this identity geometrically, define the *alternating* $\mathbb{R}$-*bilinear form* $B_\ell \in \Lambda^2(\mathfrak{g}^*)$ *associated with* ℓ *on* $\mathfrak{g}$ via the prescription (cf. Section 4.21)

$$B_\ell : \mathfrak{g} \times \mathfrak{g} \ni (X,Y) \rightsquigarrow \langle [X,Y],\ell\rangle \in \mathbb{R}.$$

Let rad B_ℓ denote the *radical* of B_ℓ in $\mathfrak{g}$, i.e., the vector subspace of $\mathfrak{g}$ which is orthogonal to the whole vector space $\mathfrak{g}$ relative to B_ℓ - then we established the inclusion

$$\mathfrak{g}_\ell \subseteq \text{rad } B_\ell.$$

Conversely, the identity $B_\ell(X,Y) = 0$ for $X \in \mathfrak{g}$ and all $Y \in \mathfrak{g}$ implies $\langle Y,\ell\rangle = \langle \exp_G(-t.\mathrm{ad}_{\mathfrak{g}}X)Y,\ell\rangle$ for all $t \in \mathbb{R}$. Thus we have

$$\text{rad } B_\ell \subseteq \mathfrak{g}_\ell.$$

This proves the following

6.2 Lemma. Retain the above notations and assumptions. For all linear forms $\ell \in \mathfrak{g}^*$ the identity

$$\mathfrak{g}_\ell = \mathrm{Lie}(G_\ell) = \text{rad } B_\ell$$

holds.

Let $\mathrm{Coad}_{\mathfrak{g}} : \mathfrak{g} \to \mathfrak{gl}(\mathfrak{g}^*)$ denote the *differential* of the Lie group morphism $\mathrm{CoAd}_G : G \to \underline{\underline{GL}}(\mathfrak{g}^*)$. Then for all linear forms $\ell \in \mathfrak{g}^*$ and all pairs $(X,Y) \in \mathfrak{g} \times \mathfrak{g}$ the identities

$$\begin{aligned}
\langle Y, \mathrm{Coad}_{\mathfrak{g}}(X)\ell\rangle &= \frac{d}{dt}\Big|_{t=0} \langle Y, \mathrm{CoAd}_G(\exp_G tX)\ell\rangle \\
&= \frac{d}{dt}\Big|_{t=0} \langle \exp_G(-t.\mathrm{ad}_{\mathfrak{g}}X)Y,\ell\rangle \\
&= - \langle (\mathrm{ad}_{\mathfrak{g}}X)Y,\ell\rangle \\
&= B_\ell(Y,X)
\end{aligned}$$

hold. From these formulae we conclude that the *transpose* of the $\mathbb{R}$-linear mapping $\mathrm{ad}_{\mathfrak{g}}(X) : \mathfrak{g} \to \mathfrak{g}$ equals $-\mathrm{Coad}_{\mathfrak{g}}(X):\mathfrak{g}^* \to \mathfrak{g}^*$ for all $X \in \mathfrak{g}$. Moreover, we have the identity

$$\mathrm{rad}\, B_\ell = \{X \in \mathfrak{g} \mid \mathrm{Coad}_{\mathfrak{g}}(X)\ell = 0\}.$$

As the next step consider for an arbitrary linear form $\ell \in \mathfrak{g}^*$ the $\mathfrak{C}^\infty$-mapping $\Phi_\ell : G \to \mathfrak{g}^*$ defined via the prescription

$$\Phi_\ell : G \ni x \rightsquigarrow \mathrm{CoAd}_G(x)\ell \in \mathfrak{g}^*.$$

Since the image $\Phi_\ell(G)$ of G under Φ_ℓ is the *coadjoint orbit* $\mathrm{CoAd}_G(G)\ell \in \mathfrak{g}^*/G$ *at* $\ell \in \mathfrak{g}^*$, i.e., the orbit of ℓ under the coadjoint linear action of G on $\mathfrak{g}^*$, the mapping Φ_ℓ is called to be the *coadjoint orbit mapping defined by* $\ell \in \mathfrak{g}^*$. It follows that the coadjoint orbit $\mathrm{CoAd}_G(G)\ell$ at $\ell \in \mathfrak{g}^*$ can be identified with the homogeneous manifold G/G_ℓ and if we denote by $\pi : G \to G/G_\ell$ the canonical surjection, the bijective mapping

$$\dot{\Phi}_\ell : G/G_\ell \to \mathrm{CoAd}_G(G)\ell$$

allows to identify G/G_ℓ with $\mathrm{CoAd}_G(G)\ell \in \mathfrak{g}^*/G$ and to equip $\mathrm{CoAd}_G(G)\ell$ with the structure of a $\mathfrak{C}^\infty$-submanifold of $\mathfrak{g}^*$ in such a way that $\dot{\Phi}_\ell$ becomes a diffeomorphism. Since

$$\Phi_\ell(\exp_G tX) = \mathrm{CoAd}_G(\exp_G tX)\ell$$

holds for all $X \in \mathfrak{g}$ and $t \in \mathbb{R}$ we obtain for the differential of Φ_ℓ at the neutral element 1_G of G:

$$D\Phi_\ell(1_G) : \mathfrak{g} \ni X \rightsquigarrow \mathrm{Coad}_{\mathfrak{g}}(X)\ell \in \mathfrak{g}^*.$$

Consequently, we get for all pairs $(X,Y) \in \mathfrak{g} \times \mathfrak{g}$ the identities

$$\begin{aligned} \langle Y, D\Phi_\ell(1_G)(X)\rangle &= \langle Y, \mathrm{Coad}_{\mathfrak{g}}(X)\ell\rangle \\ &= -\langle(\mathrm{ad}_{\mathfrak{g}}X)Y, \ell\rangle \\ &= B_\ell(Y,X). \end{aligned}$$

The image $D\Phi_\ell(1_G)(\mathfrak{g})$ of $\mathfrak{g}$ under the $\mathbb{R}$-linear mapping $D\Phi_\ell(1_G):\mathfrak{g} \to \mathfrak{g}^*$ is called to be the *tangent space at the point* $\ell \in \mathfrak{g}^*$ of the coadjoint orbit

$\mathrm{CoAd}_G(G)\ell$ of G at ℓ. It follows

6.3 Lemma. Let G be a simply connected, nilpotent, real Lie group with Lie algebra $\mathfrak{g} = \mathrm{Lie}(G)$. The tangent space at the point $\ell \in \mathfrak{g}^*$ of the coadjoint orbit $\mathrm{CoAd}_G(G)\ell \in \mathfrak{g}^*/G$ of G at ℓ is isomorphic to the quotient vector space $\mathfrak{g}/\mathrm{rad}\, B_\ell$ over $\mathbb{R}$.

Let $\dot{B}_\ell$ denote the non-degenerate alternating $\mathbb{R}$-bilinear form induced by B_ℓ on the quotient $\mathfrak{g}/\mathrm{rad}\, B_\ell$. Then we have the following consequence:

Corollary. The tangent space at the point $\ell \in \mathfrak{g}^*$ of the coadjoint orbit $\mathrm{CoAd}_G(G)\ell \in \mathfrak{g}^*/G$ of G at ℓ is a symplectic real vector space with respect to $\dot{B}_\ell$. In particular, it has even dimension over $\mathbb{R}$.

6.4 Let $\mathfrak{g}$ denote a Lie algebra over the field K having $\mathfrak{g}^*$ as its K-dual vector space. Recall from Section 4.21 supra that a Lie subalgebra $\mathfrak{h}$ of $\mathfrak{g}$ is called to be *subordinate* to $\ell \in \mathfrak{g}^*$, if $\mathfrak{h}$ forms a totally isotropic vector subspace of $\mathfrak{g}$ relative to the alternating K-bilinear form $B_\ell : \mathfrak{g} \times \mathfrak{g} \ni (X,Y) \rightsquigarrow \langle [X,Y],\ell \rangle \in K$ associated with ℓ on $\mathfrak{g}$, i.e.,

$$B_\ell | (\mathfrak{h} \times \mathfrak{h}) = \langle [\mathfrak{h},\mathfrak{h}],\ell \rangle = 0.$$

If the Lie *subalgebra* $\mathfrak{h}$ of $\mathfrak{g}$ subordinate to $\ell \in \mathfrak{g}^*$ is *maximal* among the *totally isotropic vector subspaces* of $\mathfrak{g}$ relative to B_ℓ, then $\mathfrak{h}$ is called to be a K-*polarization of* $\mathfrak{g}$ *for* ℓ. In other words, if P denotes a vector subspace of $\mathfrak{g}$ such that $\mathfrak{h} \subseteq P$ and $B_\ell | (P \times P) = 0$ then we have $\mathfrak{h} = P$. In particular, each K-polarization $\mathfrak{h}$ of $\mathfrak{g}$ for $\ell \in \mathfrak{g}^*$ is subordinate to ℓ and contains therefore rad B_ℓ. Since the centre $\mathfrak{c}$ of $\mathfrak{g}$ is contained in rad B_ℓ we have the inclusions

$$\mathfrak{c} \hookrightarrow \mathrm{rad}\, B_\ell \hookrightarrow \mathfrak{h}.$$

A maximal totally isotropic vector subspace of $\mathfrak{g}$ relative to B_ℓ need not to be a K-polarization of $\mathfrak{g}$ for $\ell \in \mathfrak{g}^*$. If $\mathfrak{g}$ is finite dimensional over K, a Lie subalgebra $\mathfrak{h}$ of $\mathfrak{g}$ subordinate to $\ell \in \mathfrak{g}^*$ of maximal dimension over K is not necessarily a K-polarization of $\mathfrak{g}$ for ℓ. Moreover, it is not true that there exist K-polarizations of general Lie algebras $\mathfrak{g}$ over K for all K-linear forms $\ell \in \mathfrak{g}^*$. However, if $\mathfrak{g}$ is a nilpotent real Lie algebra, then there exist real polarizations of $\mathfrak{g}$ for arbitrary $\mathbb{R}$-linear forms ℓ on $\mathfrak{g}$ and the subalgebras $\mathfrak{h}$ of $\mathfrak{g}$ subordinate to $\ell \in \mathfrak{g}^*$ of maximal dimension over $\mathbb{R}$ are exactly the real polarizations of $\mathfrak{g}$ for ℓ. Before establishing this in Theorem 6.8 infra in a strengthened form we have to look at some generalities on alternating K-bilinear forms. In this way, a convenient formula for computing the dimensions of K-polarizations $\mathfrak{h}$ of finite dimensional Lie algebras $\mathfrak{g}$ over K for a given K-linear form $\ell \in \mathfrak{g}^*$ will follow.

6.5 Let E be a finite dimensional vector space over the field K and $B : E \times E \to K$ an alternating K-bilinear form on E. For any subset F of E let the vector subspace

$$F^{\perp} = \{x \in E | B(x,y) = 0 \textit{ for all } y \in F\}$$

of E be the *orthogonal* of F relative to B. In particular, $E^{\perp}$ is the *radical* rad B of B in E (cf. Section 6.1). Suppose that F is a vector subspace of E and define the K-linear mapping

$$f_F : E \ni x \rightsquigarrow B(x,\cdot)|F \in (F/F \cap \text{rad } B)^*.$$

Then $\text{Ker}(f_F) = F^{\perp}$. Consequently we have

$$\dim_K F - \dim_K(F \cap \text{rad } B) + \dim_K F^{\perp} = \dim_K E.$$

The vector subspace F of E is called to be *isotropic* or *coisotropic* relative to B if $F \subseteq F^{\perp}$ resp. $F^{\perp} \subseteq F$. In the isotropic case we have

$$\dim_K F \leq \frac{1}{2}(\dim_K E + \dim_K(F \cap \text{rad } B)),$$

and in the coisotropic case

$$\dim_K F \geq \frac{1}{2}(\dim_K E + \dim_K(F \cap \text{rad } B)).$$

Since $E^{\perp} = \text{rad } B$ is an isotropic vector subspace of E, a *maximal* vector subspace F among the isotropic vector subspaces of E satisfies $\text{rad } B \subseteq F$, hence $\dim_K F \leq \frac{1}{2}(\dim_K E + \dim_K(\text{rad } B))$. The maximality property of F suggests that $\dim_K F$ actually attains the upper bound.

6.6 Lemma. A vector subspace F of the finite dimensional vector space E over K is maximal among the isotropic vector subspaces of E relative to B if and only if

$$\dim_K F = \frac{1}{2}(\dim_K E + \dim_K(\text{rad } B))$$

holds.

Proof. It will be sufficient to establish that the dimension of each vector subspace F of E which is maximal among the isotropic vector subspaces of E relative to B attains the upper bound $\frac{1}{2}(\dim_K E + \dim_K(\text{rad } B))$. Let $\dot{B}$ denote the non-degenerate alternating K-bilinear form induced by B on the quotient vector space E/rad B over K. Then $(E/\text{rad } B, \dot{B})$ is a symplectic vector space over K. In particular $\dim_K(E/\text{rad } B)$ is an even positive integer. Choose a *Lagrangian* vector subspace L of E/rad B. Since L is isotropic *and* coisotropic it coincides with its orthogonal vector subspace of E/rad B relative to the symplectic form $\dot{B}$. It follows

$$\dim_K L = \frac{1}{2}\dim_K(E/\text{rad } B).$$

Denote by $\pi : E \to E/\text{rad } B$ the canonical epimorphism and let $F = \pi^{-1}(L)$ denote the preimage of L relative to π - then F is an isotropic vector subspace of E relative to B containing rad B such that

$$\begin{aligned}\dim_K F &= \frac{1}{2}\dim_K(E/\text{rad } B) + \dim_K(\text{rad } B)\\ &= \frac{1}{2}\dim_K E - \frac{1}{2}\dim_K(\text{rad } B) + \dim_K(\text{rad } B)\\ &= \frac{1}{2}(\dim_K E + \dim_K(\text{rad } B))\end{aligned}$$

holds. Since for each isotropic vector subspace P of E relative to B containing rad B the image $\pi(P)$ is isotropic in E/rad B relative to $\dot{B}$, the proof is complete. —

If we apply the preceding lemma to the case when E is a finite dimensional Lie algebra $\mathfrak{g}$ over the field K and $B \in \Lambda^2(E^*)$ is the alternating K-bilinear form $B_\ell : \mathfrak{g} \times \mathfrak{g} \ni (X,Y) \rightsquigarrow \langle [X,Y], \ell \rangle \in K$ on $\mathfrak{g}$ defined by $\ell \in \mathfrak{g}^*$, then we get the following characterization of the K-polarizations of $\mathfrak{g}$ for ℓ:

6.7 Lemma. Let $\ell \in \mathfrak{g}^*$ denote a K-linear form on the finite dimensional Lie algebra $\mathfrak{g}$ over the field K. For a Lie subalgebra $\mathfrak{h}$ of $\mathfrak{g}$ subordinate to ℓ the following conditions are pairwise equivalent:

(i) $\mathfrak{h}$ forms a K-polarization of $\mathfrak{g}$ for ℓ;

(ii) For any element $X \in \mathfrak{g}$ such that $B_\ell(X,Y) = 0$ holds for all $Y \in \mathfrak{h}$ we have $X \in \mathfrak{h}$.

(iii) The orthogonal vector subspace $\mathfrak{h}^{\perp}$ of $\mathfrak{h}$ relative to B_{ℓ} is contained in $\mathfrak{h}$.

(iv) $\dim_K \mathfrak{h} = \frac{1}{2}(\dim_K \mathfrak{g} + \dim_K(\text{rad } B_{\ell}))$.

The next step is the aforementioned existence proof for real polarizations.

Let G be a simply connected, nilpotent, real Lie group with Lie algebra $\mathfrak{g} = \text{Lie}(G)$ and $\mathfrak{h}$ a Lie subalgebra of $\mathfrak{g}$ subordinate to a $\mathbb{R}$-linear form $\ell \in \mathfrak{g}^*$. Recall from Section 4.21 supra the definition of the monomial representation $(U_{\ell,\mathfrak{h}}, \mathcal{H})$ of G acting in the complex Hilbert space $\mathcal{H}$.

6.8 Theorem. Let G be a simply connected, nilpotent, real Lie group. For each $\mathbb{R}$-linear form $\ell \in \mathfrak{g}^*$ on the nilpotent real Lie algebra $\mathfrak{g} = \text{Lie}(G)$ there exists a real polarization $\mathfrak{h}$ of $\mathfrak{g}$ for ℓ such that the monomial representation $(U_{\ell,\mathfrak{h}}, \mathcal{H})$ of G is topologically irreducible.

Proof. The proof follows by induction on $\dim_{\mathbb{R}} \mathfrak{g}$ as usual in the representation theory of nilpotent Lie groups.

If $\dim_{\mathbb{R}} \mathfrak{g} = 1$ then $\mathfrak{g}$ is a real polarization of $\mathfrak{g}$ for ℓ and $(U_{\ell,\mathfrak{g}}, \mathcal{H})$ is the one-dimensional, continuous, unitary, linear representation $(\chi_{\ell,\mathfrak{g}} \cdot \text{id}_{\mathbb{C}}, \mathbb{C})$ of $G = \mathbb{R}$.

Let $\dim_{\mathbb{R}} \mathfrak{g} = n \geqq 2$ and suppose that the theorem has been established for all simply connected, nilpotent, real Lie groups G such that $\dim_{\mathbb{R}} \text{Lie}(G) \leqq n-1$. Let $\mathfrak{c} = \mathfrak{C}_1 \mathfrak{g}$ be the centre of $\mathfrak{g}$ and in addition suppose that $\dim_{\mathbb{R}} \mathfrak{c} \geqq 2$ holds. Then the abelian ideal

$$\mathfrak{g}' = \mathfrak{c} \cap \text{Ker}(\ell)$$

of $\mathfrak{g}$ has real dimension $\geqq 1$ and ℓ induces canonically a $\mathbb{R}$-linear form $\dot{\ell}$ on the real quotient Lie algebra $\mathfrak{g}'' = \mathfrak{g}/\mathfrak{g}'$. Let $\pi'' : \mathfrak{g} \to \mathfrak{g}''$ denote the canonical Lie algebra epimorphism and G" the unique simply connected Lie group such that $\mathfrak{g}'' = \text{Lie}(G'')$. The induction hypothesis implies that there exists a real polarization $\mathfrak{h}''$ of $\mathfrak{g}''$ at $\dot{\ell}$ such that the monomial representation $(U_{\dot{\ell},\mathfrak{h}''}, \mathcal{H})$ of G" is topologically irreducible. Define $\mathfrak{h} = \pi''^{-1}(\mathfrak{h}'')$. Then the Lie subalgebra $\mathfrak{h}$ of $\mathfrak{g}$ is subordinate to ℓ. Let $X \in \mathfrak{g}$ be an arbitrary element of the orthogonal $\mathfrak{h}^{\perp}$ of $\mathfrak{h}$ in $\mathfrak{g}$ relative to B_{ℓ}. Then $\pi''(X) \in \mathfrak{g}''$ belongs to the orthogonal of $\mathfrak{h}''$ in $\mathfrak{g}''$ relative to $B_{\dot{\ell}}$. It follows $\pi''(X) \in \mathfrak{h}''$, hence $X \in \mathfrak{h}$, and therfore $\mathfrak{h}^{\perp} \subseteqq \mathfrak{h}$. We conclude that $\mathfrak{h}$ is a real polarization of $\mathfrak{g}$ for ℓ. Consider the connected Lie subgroups G' and H of G such that

$\mathfrak{g}' = \mathrm{Lie}(G')$ and $\mathfrak{h} = \mathrm{Lie}(H)$. Then G' is a closed normal subgroup of G and we have $G'' = G/G'$. Let $\pi: G \to G''$ be the canonical Lie group epimorphism. Then π restricts to a morphism π_0 of H onto the connected Lie subgroup H" of G" such that $\mathfrak{h}'' = \mathrm{Lie}(H'')$. It can be proved that H" is a simply connected, closed Lie subgroup of G". It follows from Lemma 2.16 that the monomial representations $(U_{\ell,\mathfrak{h}}, \mathcal{H})$ and $U_{\ell,\mathfrak{h}''}^{\cdot} \circ \pi, \mathcal{H}'')$ of G are unitarily isomorphic. Therefore $(U_{\ell,\mathfrak{h}}, \mathcal{H})$ is a topologically irreducible, monomial representation of G.

Finally, suppose that $\dim_{\mathbb{R}} \mathfrak{c} = 1$. According to Section 4.16 supra there exist elements $\{X_0, Y_0, T_0\}$ in $\mathfrak{g}$ such that

$$\mathfrak{g} = \mathbb{R}.X_0 \oplus \mathfrak{g}_0,$$

where $\mathfrak{g}_0 = \mathrm{Ker}(\mathrm{ad}_{\mathfrak{g}} Y_0)$ is the centralizer of $\{Y_0\}$ in $\mathfrak{g}$, $\mathfrak{c} = \mathbb{R}.T_0$, and

$$[X_0, Y_0] = T_0.$$

Let $\ell_0 = \ell | \mathfrak{g}$ and G_0 be the (unique) connected Lie subgroup of G such that $\mathfrak{g}_0 = \mathrm{Lie}(G_0)$. The inductive hypothesis implies that there exists a real polarization $\mathfrak{h}_0$ of $\mathfrak{g}_0$ at $\ell_0 \in \mathfrak{g}_0^*$ such that the monomial representation $(U_{\ell_0,\mathfrak{h}_0}, \mathcal{H}_0)$ of G_0 is topologically irreducible. According to Section 4.17 define the abelian ideal

$$\mathfrak{n}_0 = \mathbb{R}.Y_0 \oplus \mathbb{R}.T_0$$

in $\mathfrak{g}$ having the ideal $\mathfrak{g}_0$ in $\mathfrak{g}$ as its centralizer in $\mathfrak{g}$. Since $\mathfrak{n}_0$ is contained in the centre of $\mathfrak{g}_0$, the inclusion $\mathfrak{n}_0 \subseteq \mathfrak{k}_0$ holds. Let N_0 be the unique connected Lie subgroup of G such that $\mathfrak{n}_0 = \mathrm{Lie}(N_0)$. Then N_0 is a closed, normal, abelian subgroup of G properly embedded into G. By virtue of 2.9 Remark 1 we have the identity

$$U_{\ell_0,\mathfrak{h}_0}(x) = \chi_{\ell_0,\mathfrak{h}_0}(x).\mathrm{id}_{\mathcal{H}_0}$$

for all elements $x \in N_0$. Since G_0 is the stablizer of $\chi_{\ell_0,\mathfrak{h}_0}$ by Theorem 4.19, an application of Theorem 2.20 shows that the continuous, unitary, linear representation

$$(U, \mathcal{H}) = \mathrm{Ind}_{G_0}^{G}(U_{\ell_0,\mathfrak{h}_0}, \mathcal{H}_0)$$

of G is topologically irreducible. An application of the inducing in stages (Theorem 2.14) shows that the proof is complete if it can be shown that $\mathfrak{h}_0$ forms a real polarization of $\mathfrak{g}$ for ℓ. Since $\mathfrak{h}_0$ is subordinate to $\ell_0 \in \mathfrak{g}_0^*$ it is also subordinate to $\ell \in \mathfrak{g}^*$. From the inclusion $\mathfrak{n}_0 \subseteqq \mathfrak{h}_0$ we conclude that the associated orthogonal vector subspaces in $\mathfrak{g}$ relative to B_ℓ satisfy $\mathfrak{h}_0^\perp \subseteqq \mathfrak{n}_0^\perp$. We will show that $\mathfrak{n}_0^\perp = \mathfrak{g}_0$ holds. Indeed, each element $W \in \mathfrak{g}$ admits the decomposition

$$W = tX_0 + W_0$$

where $t \in \mathbb{R}$ and $W_0 \in \mathfrak{g}_0$. It follows

$$B_\ell(tX_0 + W_0, Y_0) = tB_\ell(X_0, Y_0) = t \cdot \langle T_0, \ell \rangle.$$

Therefore $W \in \mathfrak{g}$ belongs to $\mathfrak{n}_0^\perp$ if and only if $W = W_0$. Consequently

$$\mathfrak{h}_0^\perp \subseteqq \mathfrak{h}_0.$$

Thus the orthogonal vector subspace to $\mathfrak{h}_0$ in $\mathfrak{g}$ relative to B_ℓ coincides with the orthogonal vector subspace to $\mathfrak{h}_0$ in $\mathfrak{g}_0$ relative to B_{ℓ_0}. It follows

$$\mathfrak{h}_0^\perp \subseteqq \mathfrak{h}_0.$$

An application of Lemma 6.6 supra shows that $\mathfrak{h}_0$ forms a real polarization of $\mathfrak{g}$ for ℓ. —

6.9 As in the preceding section let G be a simply connected, nilpotent, real Lie group with Lie algebra $\mathfrak{g} = \text{Lie}(G)$ and $\mathfrak{h}$ a Lie subalgebra of $\mathfrak{g}$ subordinate to the $\mathbb{R}$-linear form $\ell \in \mathfrak{g}^*$. Suppose that $(U_{\ell,\mathfrak{h}}, \mathfrak{H})$ is a topologically irreducible representation of G. Let $\mathfrak{c} = \mathfrak{C}_1\mathfrak{g}$ be the centre of $\mathfrak{g}$ and $\mathfrak{c}'$ a vector subspace of $\mathfrak{c}$ such that the direct sum decomposition

$$\mathfrak{c} = (\mathfrak{h} \cap \mathfrak{c}) \oplus \mathfrak{c}'$$

holds. Suppose that $\mathfrak{h}$ does not contain $\mathfrak{c}$. Then $\dim_{\mathbb{R}} \mathfrak{c}' \geqq 1$. Let $\mathfrak{h}' = \mathfrak{c}' \oplus \mathfrak{h}$ and denote by H, H' and C' the connected Lie subgroups of G having the Lie algebras $\mathfrak{h}$, $\mathfrak{h}'$ and $\mathfrak{c}'$, respectively. Then H' is the semi-direct product of C' and H. The unitarily induced representation

$$\mathrm{Ind}_H^{H'}(\chi_{\ell,\mathfrak{h}}\cdot\mathrm{id}_{\mathbb{C}},\mathbb{C})$$

is topologically reducible since it can be realized on the complex Hilbert space $L^2(C')$ and the restrictions to C' and H are the regular representation of C' and a multiple of $(\chi_{\ell,\mathfrak{h}}\cdot\mathrm{id}_{\mathbb{C}},\mathbb{C})$, respectively. It follows by Remark 2 of Section 2.9 that

$$(U_{\ell,\mathfrak{h}},\mathcal{H}) = \mathrm{Ind}_{H'}^{G}(\mathrm{Ind}_H^{H'}(\chi_{\ell,\mathfrak{h}}\cdot\mathrm{id}_{\mathbb{C}},\mathbb{C}))$$

is not topologically irreducible which contradicts the assumption. Thus $\mathfrak{c} \subseteq \mathfrak{h}$ and we have proved the following

6.10 Lemma. Let G be a simply connected, nilpotent, real Lie group with Lie algebra $\mathfrak{g}$ and $\ell \in \mathfrak{g}^*$ a $\mathbb{R}$-linear form on $\mathfrak{g}$. Any Lie subalgebra $\mathfrak{h}$ of $\mathfrak{g}$ subordinate to ℓ such that the monomial representation $(U_{\ell,\mathfrak{h}},\mathcal{H})$ of G is topologically irreducible, contains the centre $\mathfrak{c}$ of $\mathfrak{g}$.

6.11 Keeping to the above notations and assumptions, define the abelian ideal

$$\mathfrak{g}' = \mathfrak{c} \cap \mathrm{Ker}(\ell)$$

of $\mathfrak{g}$ with real dimension

$$\dim_{\mathbb{R}}\mathfrak{g}' \geqq \dim_{\mathbb{R}}\mathfrak{c} - 1.$$

In the case $\dim_{\mathbb{R}}\mathfrak{g}' \geqq 1$ the linear form $\ell \in \mathfrak{g}^*$ induces canonically a $\mathbb{R}$-linear form $\dot{\ell}$ on the real quotient Lie algebra $\mathfrak{g}'' = \mathfrak{g}/\mathfrak{g}'$. Let $\pi'': \mathfrak{g} \to \mathfrak{g}''$ denote the canonical Lie algebra epimorphism and $\mathfrak{h}'' = \pi''(\mathfrak{h})$. Then $\mathfrak{h}''$ is subordinate to $\dot{\ell} \in (\mathfrak{g}'')^*$ and if $\mathfrak{h}$ forms a real polarization of $\mathfrak{g}$ for ℓ then $\mathfrak{h}''$ forms a real polarization of $\mathfrak{g}''$ for $\dot{\ell}$. Conversely, if $\mathfrak{h}''$ is a real polarization of $\mathfrak{g}''$ for $\dot{\ell}$, its preimage $\pi^{-1}(\mathfrak{h}'') = \mathfrak{h}'' + \mathfrak{g}'$ is a real polarization of $\mathfrak{g}$ for ℓ. In view of $\mathfrak{g}' \subseteq \mathfrak{c} \subseteq \mathfrak{h}$ by Lemma 6.10, it follows that $\mathfrak{h}$ is a real polarization of $\mathfrak{g}$ for ℓ if and only if $\mathfrak{h}''$ forms a real polarization of $\mathfrak{g}''$ for $\dot{\ell}$. If G' denotes the connected Lie subgroup of G such that $\mathfrak{g}' = \mathrm{Lie}(G)$, then G' is a closed normal subgroup of G and the quotient group G" = G/G' has the Lie algebra $\mathfrak{g}''$. Let $\pi: G \to G''$ denote the corresponding canonical epimorphism at the Lie group level. An application of Lemma 2.16 shows that

$$U_{\ell,\mathfrak{h}} = U_{\dot{\ell},\mathfrak{h}''} \circ \pi$$

holds. It follows by induction on $\dim_{\mathbb{R}}\mathfrak{g}$ that $(U_{\ell,\mathfrak{h}},\mathcal{H})$ is a topologically irreducible representation of G if and only if $\mathfrak{h}$ is a real polarization of $\mathfrak{g}$ for ℓ. This holds in the easy case $\dim_{\mathbb{R}}\mathfrak{g}' \geqq 1$. In the case $\dim_{\mathbb{R}}\mathfrak{g}' = 0$, however, a more refined analysis based on the Mackey machinery shows that the result continues to hold. To summarize, we have the following result:

6.12 Theorem. Let G be a simply connected, nilpotent, real Lie group with Lie algebra $\mathfrak{g} = \mathrm{Lie}(G)$ and $\mathfrak{h}$ a Lie subalgebra of $\mathfrak{g}$ subordinate to the $\mathbb{R}$-linear form $\ell \in \mathfrak{g}^*$. The monomial representation $(U_{\ell,\mathfrak{h}},\mathcal{H})$ of G is topologically irreducible if and only if $\mathfrak{h}$ forms a real polarization of $\mathfrak{g}$ for ℓ.

Proof. The theorem is clear for $\dim_{\mathbb{R}}\mathfrak{g} = 1$. In view of the reasonings of the preceding section we may apply induction on $\dim_{\mathbb{R}}\mathfrak{g}$ by assuming $\dim_{\mathbb{R}}\mathfrak{g}' = 0$. This means that $\mathfrak{c} = \mathbb{R}.T$ is of real dimension 1, and $\ell|\mathfrak{c} \neq 0$. Choose the vector $T \in \mathfrak{c}$ such that $\langle T,\ell\rangle = 1$. Choose an abelian ideal $\mathfrak{a}$ in $\mathfrak{g}$ such that $\mathfrak{c} \subseteqq \mathfrak{a}$ and $\dim_{\mathbb{R}}\mathfrak{a} = 2$. This is always possible as $\mathfrak{g}$ is nilpotent, so we need only choose $Y \in \mathfrak{C}^2\mathfrak{g}$ such that $Y \notin \mathfrak{c}$ and write

$$\mathfrak{a} = \mathbb{R}.Y \oplus \mathbb{R}.T$$

where $\langle Y,\ell\rangle = 0$. Indeed, we only need to choose $Y \in \mathfrak{a} \cap \mathrm{Ker}(\ell)$ with $Y \neq 0$, as $\mathrm{Ker}(\ell)$ has real codimension 1 and $\mathfrak{c} \cap \mathrm{Ker}(\ell) = \{0\}$. Look at the centralizer $\mathfrak{g}_0$ of $\mathfrak{a}$ in $\mathfrak{g}$. Then $\mathfrak{g}_0$ is an ideal in $\mathfrak{g}$ of real codimension 1. This is because we only need to look at

$$[X,Y] = \lambda(X).T$$

as $[X,T] = 0$ for all $X \in \mathfrak{g}$. Then $\lambda \neq 0$ is a $\mathbb{R}$-linear form on $\mathfrak{g}$ having $\mathfrak{g}_0$ as its kernel. It follows $\mathrm{codim}_{\mathbb{R}}\mathfrak{g}_0 = 1$ as stated and the Jacobi identity yields for $X \in \mathfrak{g}$ and $X_0 \in \mathfrak{g}_0$ the equalities

$$\begin{aligned}[][[X,X_0],Y] &= [[X_0,Y],X] - [[Y,X],X_0] \\ &= \lambda(X)[T,X_0] \\ &= 0.\end{aligned}$$

Hence $[X,X_0] \in \mathfrak{g}_0$.

Let $\mathfrak{h}$ be a Lie subalgebra of $\mathfrak{g}$ subordinate to ℓ such that

$$\mathfrak{c} \hookrightarrow \mathfrak{h} \hookrightarrow \mathfrak{g}_0$$

and $\ell_0 = \ell|\mathfrak{g}_0$. Obviously $\mathfrak{h}$ is subordinate to ℓ_0 and a real polarization of $\mathfrak{g}_0$ for ℓ_0 if $\mathfrak{h}$ is a real polarization of $\mathfrak{g}$ for ℓ.

Conversely, if $\mathfrak{h}$ is a real polarization of $\mathfrak{g}_0$ for ℓ_0 then $\mathfrak{a} \subseteq \mathfrak{h}$. It follows $\mathfrak{h}^\perp \subseteq \mathfrak{a}^\perp = \mathfrak{g}_0$. We conclude that $\mathfrak{h}$ forms a real polarization of $\mathfrak{g}$ for ℓ if and only if $\mathfrak{h}$ forms a real polarization of $\mathfrak{g}_0$ for ℓ_0.

Let $H = \exp_G \mathfrak{h}$ and $G_0 = \exp_G \mathfrak{g}_0$. Then we have

$$\begin{aligned}(U_{\ell,\mathfrak{h}}, \mathcal{H}) &= \mathrm{Ind}_H^G(\chi_{\ell,\mathfrak{h}} \cdot \mathrm{id}_{\mathbb{C}}, \mathbb{C}) \\ &= \mathrm{Ind}_{G_0}^G(\mathrm{Ind}_H^{G_0}(\chi_{\ell,\mathfrak{h}} \cdot \mathrm{id}_{\mathbb{C}}, \mathbb{C})) \\ &= \mathrm{Ind}_{G_0}^G(U_{\ell_0,\mathfrak{h}}, \mathcal{H}_0)\end{aligned}$$

and $(U_{\ell_0,\mathfrak{h}}, \mathcal{H}_0)$ is topologically irreducible when $(U_{\ell,\mathfrak{h}}, \mathcal{H})$ has this property. Conversely, G_0 is the stabilizer of $\chi_{\ell,\mathfrak{h}}$ in G by Theorem 4.19, and if we restrict $U_{\ell_0,\mathfrak{h}}$ to the analytic, normal, abelian subgroup $A = \exp_G \mathfrak{a}$ of G corresponding to the abelian ideal $\mathfrak{a}$ of $\mathfrak{g}$ it is a multiple of $\chi_{\ell,\mathfrak{h}}$. It follows by Theorem 2.20 that $(U_{\ell,\mathfrak{h}}, \mathcal{H})$ is topologically irreducible if and only if $(U_{\ell_0,\mathfrak{h}}, \mathcal{H}_0)$ is topologically irreducible.

Now suppose that the Lie subalgebra $\mathfrak{h}$ of $\mathfrak{g}$ subordinate to ℓ satisfies $\mathfrak{c} \subseteq \mathfrak{h}$ and $\mathfrak{h} \not\subseteq \mathfrak{g}_0$. It follows $\mathfrak{a} \not\subseteq \mathfrak{h}$ since otherwise $\mathfrak{h} \subseteq \mathfrak{h}^\perp \subseteq \mathfrak{a}^\perp = \mathfrak{g}_0$. Consequently $Y \notin \mathfrak{h}$. We can choose $X \in \mathfrak{h}$ such that

$$\langle X, \lambda\rangle = 1, \quad \langle X, \ell\rangle = 0.$$

It follows $\mathfrak{g} = \mathbb{R}.X \oplus \mathfrak{g}_0$ and $[X,Y] = T$. Put $\mathfrak{h}_0 = \mathfrak{h} \cap \mathfrak{g}_0$, $\mathfrak{h}' = \mathfrak{h}_0 + \mathfrak{a}$, and $\mathfrak{k} = \mathfrak{h} + \mathfrak{a}$. Then the proof of Theorem 6.12 will follow from the next two lemmas.

6.13 Lemma. Keeping to the notations above we have

(i) $\mathfrak{h} = \mathbb{R}.X \oplus \mathfrak{h}_0$;

(ii) $\mathfrak{h}' = \mathbb{R}.Y \oplus \mathfrak{h}_0$ and $\mathfrak{h}'$ satisfies $\langle [X', \mathfrak{h}'], \ell\rangle = 0$ if and only if $X' \in \mathfrak{h}'$;

(iii) $\mathfrak{k} = \mathbb{R}.X \oplus \mathbb{R}.Y \oplus \mathfrak{h}_0$ is a Lie subalgebra of $\mathfrak{g}$ and every element in $K = \exp_G \mathfrak{k}$ can be written in a unique way as $(\exp_G uX)h_0(\exp_G vY)$ where $h_0 \in \exp_G \mathfrak{h}_0$ and $(u,v) \in \mathbb{R} \oplus \mathbb{R}$.

Proof.

(i) Obviously $\mathfrak{h}_0$ is of real codimension 1 in $\mathfrak{h}$ and $X \notin \mathfrak{g}_0$.

(ii) Since $\mathfrak{a}$ is an abelian ideal in $\mathfrak{g}$ we have $\langle[\mathfrak{a},\mathfrak{a}],\ell\rangle = 0$. Moreover $\mathfrak{h}_0 \subseteq \mathfrak{h}$ and $\mathfrak{h}$ is a totally isotropic vector subspace of $\mathfrak{g}$ relative to B_ℓ. Thus $\langle[\mathfrak{h}_0,\mathfrak{h}_0],\ell\rangle = 0$. Finally $\mathfrak{h}_0 \subseteq \mathfrak{g}_0$ implies $[\mathfrak{h}_0,\mathfrak{a}] = 0$ and therefore

$$\begin{aligned}\langle[\mathfrak{h}',\mathfrak{h}'],\ell\rangle &= \langle[\mathfrak{h}_0,\mathfrak{h}_0],\ell\rangle + \langle[\mathfrak{h}_0,\mathfrak{a}],\ell\rangle + \langle[\mathfrak{a},\mathfrak{a}],\ell\rangle \\ &= 0.\end{aligned}$$

Hence $\mathfrak{h}'$ is a totally isotropic vector subspace of $\mathfrak{g}$ relative to B_ℓ. But $Y \notin \mathfrak{h}_0$ because $Y \notin \mathfrak{h}$, so $\mathfrak{h}' \supseteq \mathbb{R}.Y \oplus \mathfrak{h}_0$. But $\dim_{\mathbb{R}}(\mathbb{R}.Y \oplus \mathfrak{h}_0) = \dim_{\mathbb{R}}\mathfrak{h}$. Since all maximal totally isotropic vector subspaces of $\mathfrak{g}$ with respect to the given $\mathbb{R}$-bilinear form B_ℓ have the same real dimension we conclude that $\dim_{\mathbb{R}}\mathfrak{h}' \leq \dim_{\mathbb{R}}\mathfrak{h}$. Thus we have $\dim_{\mathbb{R}}\mathfrak{h}' = \dim_{\mathbb{R}}\mathfrak{h}$ and so $\mathfrak{h}' = \mathbb{R}.Y \oplus \mathfrak{h}_0$ is a maximal totally isotropic vector subspace of $\mathfrak{g}$ relative to B_ℓ.

(iii) Observe that $\mathbb{R}.Y \oplus \mathfrak{h}_0$ is an ideal in $\mathfrak{k}$ as

$$[X,\mathfrak{h}_0] \subseteq [\mathfrak{h},\mathfrak{h}] \cap \mathfrak{g}_0 \subseteq \mathfrak{h}_0, \qquad [X,Y] = T \in \mathfrak{h}_0,$$

so $\mathfrak{k} = \mathbb{R}.X \oplus (\mathfrak{h}_0 \oplus \mathbb{R}.Y)$ is a Lie subalgebra of $\mathfrak{g}$ and any element of K can be written in a unique way as $(\exp_G uX).(\exp_G Y_0)$ where $Y_0 \in (\mathfrak{h}_0 \oplus \mathbb{R}.Y)$. But $[\mathfrak{h}_0,Y] = 0$ as $\mathfrak{h}_0 \subseteq \mathfrak{g}_0$, so the element is of the form $(\exp_G uX).h_0.(\exp_G vY)$ where $h_0 \in \exp_G \mathfrak{h}_0$ and $(u,v) \in \mathbb{R} \oplus \mathbb{R}$. —

6.14 Lemma. The monomial representations $(U_{\ell,\mathfrak{h}},\mathcal{H})$ and $(U_{\ell,\mathfrak{h}'},\mathcal{H}')$ of G are unitarily isomorphic.

Proof. It is sufficient to prove that $(U_{\ell,\mathfrak{h}},\mathcal{H})$ and $(U_{\ell,\mathfrak{h}'},\mathcal{H}')$ define unitarily isomorphic linear representations of K by Theorem 2.14 on induction in stages. Let $H' = \exp_G \mathfrak{h}'$ be the analytic subgroup of G with Lie algebra $\mathfrak{h}'$. Then $k \in K$ is uniquely decomposed by

$$k = (\exp_G yY).h \qquad (h \in H),$$

$$= (\exp_G xX).h' \qquad (h' \in H'),$$

where $y \in \mathbb{R}$ and $x \in \mathbb{R}$. The $\mathbb{R}$-linear mappings

$$\mathcal{K}(K/H, \chi_{\ell,\mathfrak{h}}) \ni \phi \rightsquigarrow (y \rightsquigarrow (\exp_G yY)) \in \mathcal{K}(\mathbb{R}),$$

$$\mathcal{K}(K/H', \chi_{\ell,\mathfrak{h}'}) \ni \phi \rightsquigarrow (x \rightsquigarrow (\exp_G xX)) \in \mathcal{K}(\mathbb{R})$$

extend to unitary isomorphisms of the complex Hilbert spaces $L^2(K/H, \chi_{\ell,\mathfrak{h}};\mu)$ and $L^2(K/H', \chi_{\ell,\mathfrak{h}'};\mu)$ respectively, onto the complex Hilbert space $L^2(\mathbb{R})$. For $\phi \in \mathcal{K}(K/H, \chi_{\ell,\mathfrak{h}})$ and $x \in \mathbb{R}$ put

$$\begin{aligned} T_{\mathfrak{h},\mathfrak{h}'}(\phi)(\exp_G xX) &= \int_{\mathbb{R}} \phi(\exp_G xX.\exp_G vY)dv \\ &= \int_{\mathbb{R}} \phi(\exp_G vY.\exp_G x(\exp_G(-v.\mathrm{ad}_{\mathfrak{g}} Y)X))dv \\ &= \int_{\mathbb{R}} \phi(\exp_G vY.\exp_G xX.\exp_G vxT)dv. \end{aligned}$$

Now $X \in \mathfrak{h}$ and $T \in \mathfrak{h}$ and the covariance formula yields

$$\begin{aligned} \bar{\chi}_{\ell,\mathfrak{h}}(\exp_G xX.\exp_G vxT) &= \bar{\chi}_{\ell,\mathfrak{h}}(\exp_G xX)\bar{\chi}_{\ell,\mathfrak{h}}(\exp_G vyT) \\ &= e^{-2\pi i x\langle X,\ell\rangle} \,.\, e^{-2\pi i vx\langle T,\ell\rangle} \end{aligned}$$

for $x \in \mathbb{R}$ and $v \in \mathbb{R}$. Since $\langle X,\ell\rangle = 0$ and $\langle T,\ell\rangle = 1$ we get

$$T_{\mathfrak{h},\mathfrak{h}'}(\phi)(\exp_G xX) = \int_{\mathbb{R}} \phi(\exp_G vY)e^{-2\pi i xv}dv.$$

Thus $T_{\mathfrak{h},\mathfrak{h}'}(\phi)(\exp_G xX)$ is the Fourier transform of the function $v \rightsquigarrow (\exp_G vY)$ at $\exp_G xX$. Consequently there exists a unique unitary isomorphism $\tilde{T}_{\mathfrak{h},\mathfrak{h}'} : L^2(K/H, \chi_{\ell,\mathfrak{h}};\mu) \to L^2(K/H', \chi_{\ell,\mathfrak{h}'};\mu)$ which intertwines the monomial representations $(U_{\ell,\mathfrak{h}}, \mathcal{H})$ and $(U_{\ell,\mathfrak{h}'}, \mathcal{H}')$ of K such that the following diagram is commutative:

$$\begin{array}{ccc} L^2(K/H,\chi_{\ell,\mathfrak{h}};\mu) & \xrightarrow{T_{\mathfrak{h},\mathfrak{h}'}} & L^2(K/H',\chi_{\ell,\mathfrak{h}'};\mu) \\ \downarrow & & \downarrow \\ L^2(\mathbb{R}) & \xrightarrow{\tilde{T}_{\mathfrak{h},\mathfrak{h}'}} & L^2(\mathbb{R}) \end{array}$$

This gives the unitary isomorphy of the monomial representations $(U_{\ell,\mathfrak{h}},\mathcal{H})$ and $(U_{\ell,\mathfrak{h}'},\mathcal{H}')$ of K and therefore proves the lemma.

6.15 Let us agree to retain the notations employed in Theorem 6.12 supra - thus, in particular, $\mathfrak{h}$ is a real polarization of $\mathfrak{g}$ for $\ell \in \mathfrak{g}^*$. Let $\mathfrak{h}^\perp$ be the orthogonal of $\mathfrak{h}$ in $\mathfrak{g}^*$ and $\ell + \mathfrak{h}^\perp$ the affine subspace of $\mathfrak{g}^*$ through ℓ of direction $\mathfrak{h}^\perp$. It follows that $\mathfrak{h}$ is also a real polarization of $\mathfrak{g}$ for each linear form $\ell_0 \in \ell + \mathfrak{h}^\perp$. Indeed, the Lie subalgebra $\mathfrak{h}$ of $\mathfrak{g}$ is subordinate to ℓ_0 and in view of $\ell|\mathfrak{h} = \ell_0|\mathfrak{h}$ we have the identity

$$(U_{\ell,\mathfrak{h}},\mathcal{H}) = (U_{\ell_0,\mathfrak{h}},\mathcal{H}).$$

Consequently the monomial representation $(U_{\ell_0,\mathfrak{h}},\mathcal{H})$ of G is topologically irreducible by Theorem 6.12 and therefore $\mathfrak{h}$ forms a real polarization of $\mathfrak{g}$ for $\ell_0 \in \ell + \mathfrak{h}^\perp$.

6.16 Lemma. Let H be the connected Lie subgroup of G such that $\mathfrak{h} = \mathrm{Lie}(H)$ - then

$$\mathrm{CoAd}_G(H)\ell = \ell + \mathfrak{h}^\perp$$

for all $\mathbb{R}$-linear forms $\ell \in \mathfrak{g}^*$.

Proof. Since $\mathfrak{h}$ is subordinate to $\ell \in \mathfrak{g}^*$ we have

$$\mathrm{Coad}_{\mathfrak{g}}(\mathfrak{h})\ell \subseteq \mathfrak{h}^\perp.$$

On the other hand, $\mathfrak{h}^\perp$ is stable under the coadjoint action of $\mathfrak{h}$, i.e., we have the inclusion

$$\mathrm{Coad}_{\mathfrak{g}}(\mathfrak{h})\mathfrak{h} \subseteq \mathfrak{h}^\perp.$$

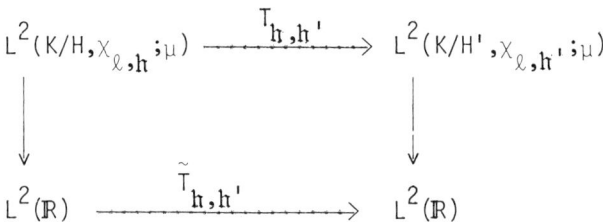

Consequently

$$\mathrm{Coad}_{\mathfrak{g}}(\mathfrak{h})(\ell+\mathfrak{h}^{\perp}) \subseteq \mathfrak{h}^{\perp}.$$

An application of the global diffeomorphism $\exp_G|H = \exp_H: \mathfrak{h} \to H$ shows that for each $\mathbb{R}$-linear form $\ell_0 \in \ell + \mathfrak{h}^{\perp}$ the inclusion

$$\mathrm{CoAd}_G(H)\ell_0 \subseteq \ell + \mathfrak{h}^{\perp}$$

holds. As in Section 6.2 let Φ_0 denote the coadjoint orbit mapping defined by ℓ_0. Then the $\mathbb{R}$-linear mapping $D\Phi_0(1_G)|\mathfrak{h}$ has the kernel $\mathfrak{h} \cap \mathrm{rad}\, B_{\ell_0}$. Since $\mathfrak{h}$ is a real polarization of $\mathfrak{g}$ for $\ell_0 \in \ell + \mathfrak{h}^{\perp}$ by Section 6.15, we have $\mathrm{rad}\, B_{\ell_0} \subseteq \mathfrak{h}$ and therefore the real dimension of the tangent space at the point ℓ_0 of the coadjoint orbit $\mathrm{CoAd}_G(H)\ell_0$ of H in $\mathfrak{g}^*$ at ℓ_0 is by Lemma 6.7

$$\dim_{\mathbb{R}}\mathfrak{h} - \dim_{\mathbb{R}}(\mathrm{rad}\, B_{\ell_0}) = \dim_{\mathbb{R}}\mathfrak{g} - \dim_{\mathbb{R}}\mathfrak{h} = \dim_{\mathbb{R}}\mathfrak{h}^{\perp}.$$

Therefore the set of coadjoint orbits

$$\{\mathrm{CoAd}_G(H)\ell_0 \,|\, \ell_0 \in \ell + \mathfrak{h}^{\perp}\}$$

is open in $\ell + \mathfrak{h}^{\perp}$. It follows that the specific coadjoint orbit $\mathrm{CoAd}_G(H)\ell$ is an open and closed subset of $\ell + \mathfrak{h}^{\perp}$. Since $\ell + \mathfrak{h}^{\perp}$ is connected, the result follows. —

Since $\mathfrak{h}$ is a real polarization of $\mathfrak{g}$ for $\ell \in \mathfrak{g}^*$ we have $\mathfrak{g}_{\ell} \subseteq \mathfrak{h}$. By virtue of $G_{\ell} = \exp_G(\mathfrak{g}_{\ell})$, G_{ℓ} is a connected Lie subgroup of G. It follows $G_{\ell} \subseteq H$ and by 6.2 that H/G_{ℓ} may be identified with $\ell + \mathfrak{h}^{\perp}$. Using this identification, the Lebesgue measure $d\ell'$ on $\mathfrak{h}^{\perp}$ gives rise to a measure on the homogeneous manifold H/G_{ℓ} which is invariant under the action of H.

6.17 <u>Lemma</u>. The measure

$$\alpha: \mathcal{K}(\ell+\mathfrak{h}^{\perp}) \ni f \rightsquigarrow \int_{\mathfrak{h}^{\perp}} f(\ell+\ell')d\ell'$$

on the homogeneous manifold H/G_{ℓ} is invariant under the coadjoint action of the group H.

<u>Proof</u>. For all elements $y \in H$ we have $\mathrm{CoAd}_G(y)\ell - \ell \in \mathfrak{h}^{\perp}$ and therefore we get for all functions $f \in \mathcal{K}(\ell+\mathfrak{h}^{\perp})$ the identity

$$\int_{\mathfrak{h}^\perp} f(\mathrm{CoAd}_G(y)(\ell+\ell'))d\ell' = \int_{\mathfrak{h}^\perp} f(\ell+\mathrm{CoAd}_G(y)\ell')d\ell'.$$

In view of $|\det(\mathrm{CoAd}_G(y^{-1})|\mathfrak{h}^\perp)| = 1$ for all $y \in H$ (cf. Section 4.14), a change of variables yields the result. —

6.18 Since the simply connected, nilpotent, real Lie groups G are exponential groups by virtue of Theorem 4.12, the global diffeomorphism $\exp_G$ of $\mathfrak{g} = \mathrm{Lie}(G)$ onto G allows one *to relate abelian harmonic analysis on the additive, locally compact, topological group* $\mathfrak{g}$ *to harmonic analysis on* G. Let the (left and right) Haar measure dx on G be the image of Lebesgue measure dX of $\mathfrak{g}$ under $\exp_G$ - then

$$\int_G \phi(x)dx = \int_{\mathfrak{g}} \phi(\exp_G X)dX$$

holds for all functions $\phi \in K(G)$ (cf. Section 4.14). If $\mathfrak{h}$ denotes a polarization of $\mathfrak{g}$ for $\ell \in \mathfrak{g}^*$ and H the connected Lie subgroup of G such that $\mathfrak{h} = \mathrm{Lie}(H)$, there exists a Haar measure dh on H such that the identity

$$\int_H \psi(h)dh = \int_{\mathfrak{h}} \psi(\exp_H Y)dY$$

holds for all functions $\psi \in K(H)$. The quotient measure dX/dY is a Haar measure on the additive, locally compact, topolgical group $\mathfrak{g}$, the vector space dual $(\mathfrak{g}/\mathfrak{h})^*$ of which identifies with the orthogonal $\mathfrak{h}^\perp$ of $\mathfrak{k}$ in $\mathfrak{g}$. Let $d\ell$ be the Haar measure on $\mathfrak{h}^\perp$ associated with dX/dY. If $f \in \mathcal{D}(\mathfrak{g})$ denotes a complex-valued, infinitely differentiable, compactly supported function on $\mathfrak{g}$ with Fourier cotransform

$$\bar{\mathcal{F}}_{\mathfrak{g}} f : \mathfrak{g}^* \ni \ell' \rightsquigarrow \int_{\mathfrak{g}} f(X)e^{2\pi i\langle X,\ell'\rangle}dX,$$

then the *Poisson formula* reads as follows:

$$\int_{\mathfrak{h}} f(Y)dY = \int_{\mathfrak{h}^\perp} (\bar{\mathcal{F}}_{\mathfrak{g}} f)(\ell')d\ell'$$

The preceding formula presents an important step in the proof of the Kirillov character formula.

6.19 Theorem (Kirillov). Let G be a simply connected, nilpotent, real Lie group with Lie algebra $\mathfrak{g}$ and $\mathfrak{h}$ a real polarization of $\mathfrak{g}$ for the $\mathbb{R}$-linear form

$\ell \in \mathfrak{g}^*$. Then the monomial representation $(U_{\ell,\mathfrak{h}}, \mathcal{H})$ of G extends to an operator of trace class

$$U^1_{\ell,\mathfrak{h}}(\phi) : \mathcal{H} \to \mathcal{H}$$

for all functions $\phi \in \mathcal{D}(G)$. There exists a measure β on the coadjoint orbit $CoAd_G(G)\ell = G/G_\ell$ which is invariant under the coadjoint action of G such that the character formula

$$\mathrm{tr}\, U^1_{\ell,\mathfrak{h}}(\phi) = \int_{CoAd_G(G)\ell} \bar{\mathcal{F}}_{\mathfrak{g}}(\phi \circ \exp_G)(\ell')d\beta(\ell')$$

holds for all test functions $\phi \in \mathcal{D}(G)$.

<u>Proof</u>. Let the measure α on the homogeneous manifold H/G_ℓ be defined according to Lemma 6.17 and define the quotient measure $\mu = \frac{dx}{dh}$ on G/H. In view of Theorem 2.7 the measure

$$\beta = \int_{G/H} (CoAd_G(x)\alpha)d\mu(\dot{x})$$

on the coadjoint orbit $CoAd_G(G)\ell = G/G_\ell$ is invariant under the coadjoint action of G and we have

$$\beta : \mathcal{K}(\mathfrak{g}^*) \ni \Phi \rightsquigarrow \int_{G/H} \left(\int_{\mathfrak{h}} \Phi(CoAd_G(x)(\ell+\ell'))d\ell'\right)d\mu(\dot{x}).$$

Let $K_\phi : (G/H) \times (G/H) \to \mathbb{C}$ denote the kernel of the continuous linear mapping $U^1_{\ell,\mathfrak{h}}(\phi): \mathcal{H} \to \mathcal{H}$ (Section 2.9, Remark 3). Then we have (cf. Section 4.21)

$$\begin{aligned} K_\phi(x,x) &= \int_H ((\check{\gamma}_G(x) \times \check{\delta}_G(x))\phi(h)).\chi_{\ell,\mathfrak{h}}(h)dh \\ &= \int_{\mathfrak{h}} \phi(x(\exp_G Y)x^{-1})e^{2\pi i\langle Y,\ell\rangle}dY \\ &= \int_{\mathfrak{h}} (\phi \circ \exp_G)(Ad_G(x)Y)e^{2\pi i\langle Y,\ell\rangle}dY \end{aligned}$$

for $x \in G$. An application of the Poisson formula (cf. 6.18 supra) gives

$$K_\phi(x,x) = \int_{\mathfrak{h}^\perp} (\bar{\mathcal{F}}_{\mathfrak{g}}(\phi \circ \exp_G)(\mathrm{Ad}_G(x)Y).e^{2\pi i\langle Y,\ell\rangle})(\ell')d\ell'$$

$$= \int_{\mathfrak{h}^\perp}(\int_{\mathfrak{g}} (\phi \circ \exp_G)(\mathrm{Ad}_G(x)X))e^{2\pi i\langle X,\ell+\ell'\rangle}dX)d\ell'$$

$$= \int_{\mathfrak{h}^\perp} (\bar{\mathcal{F}}_{\mathfrak{g}}(\phi \circ \exp_G)(\mathrm{CoAd}_G(x)(\ell+\ell'))d\ell'.$$

It can be established that β is a tempered measure on $\mathfrak{g}^*$. Therefore the function $\bar{\mathcal{F}}_{\mathfrak{g}}(\phi \circ \exp_G)$ is integrable over $\mathfrak{g}^*$ with respect to β. Hence $G/H \ni \dot{x} \rightsquigarrow K_\phi(\dot{x},\dot{x})$ is integrable with respect to μ. It follows by Mercer's theorem that the positive Hermitian operator

$$\phi \rightsquigarrow U^1_{\ell,\mathfrak{h}}(\phi) \circ U^1_{\ell,\mathfrak{h}}(\phi)^* = U^1_{\ell,\mathfrak{h}}(\phi * \check{\bar{\phi}})$$

is of trace class and therefore $U^1_{\ell,\mathfrak{h}}(\phi)$ is a Hilbert-Schmidt operator in $\mathcal{H}$ for all $\phi \in \mathcal{D}(G)$. Thus $U^1_{\ell,\mathfrak{h}}(\phi)$ is itself of trace class. Moreover, we have

$$\mathrm{tr}\, U^1_{\ell,\mathfrak{h}}(\phi) = \int_{G/H} K_\phi(x,x)d\mu(\dot{x})$$

$$= \int_{\mathrm{CoAd}_G(G)\ell} \bar{\mathcal{F}}_{\mathfrak{g}}(\phi \circ \exp_G)(\ell')d\beta(\ell')$$

for all test functions $\phi \in \mathcal{D}(G)$, and this is the desired character formula.—

Corollary. Let $\mathfrak{h}_1,\mathfrak{h}_2$ denote real polarizations of $\mathfrak{g}$ for the same linear form $\ell \in \mathfrak{g}^*$. Then the monomial representations $(U_{\ell,\mathfrak{h}_1},\mathcal{H}_1)$ and $(U_{\ell,\mathfrak{h}_2},\mathcal{H}_2)$ of G are unitarily isomorphic.

Proof. There are two measures β_1 and β_2 on the coadjoint orbit $\mathrm{CoAd}_G(G)\ell = G/G_\ell$ invariant under the coadjoint action of G such that

$$\mathrm{tr}\, U^1_{\ell,\mathfrak{h}_j}(\phi) = \int_{\mathrm{CoAd}_G(G)\ell} \bar{\mathcal{F}}_{\mathfrak{g}}(\phi \circ \exp_G)(\ell')d\beta_j(\ell')$$

holds for all test functions $\phi \in \mathcal{D}(G)$ and $j \in \{1,2\}$. Since we have $\beta_2 = \rho\beta_1$ with a real number $\rho > 0$, we obtain

$$\mathrm{tr}\, U^1_{\ell,\mathfrak{h}_2}(\phi) = \rho.\mathrm{tr}\, U^1_{\ell,\mathfrak{h}_1}(\phi)$$

for $\phi \in \mathcal{D}(G)$. An application of Theorems 6.12 and 3.10 yields $\rho = 1$ and the existence of an unitary isomorphism of $(U_{\ell,\mathfrak{h}_1}, \mathcal{H}_1)$ onto $(U_{\ell,\mathfrak{h}_2}, \mathcal{H}_2)$. —

6.20 Theorem. Let G be a simply connected, nilpotent, real Lie group with Lie algebra $\mathfrak{g}$ and ℓ_1, ℓ_2 two $\mathbb{R}$-linear forms belonging to $\mathfrak{g}^*$. If $\mathfrak{h}_j$ denotes a real polarization of $\mathfrak{g}$ for ℓ_j, the monomial representations $(U_{\ell_j,\mathfrak{h}_j}, \mathcal{H}_j)$ of G, $j \in \{1,2\}$, are unitarily isomorphic if and only if

$$\ell_2 \in \mathrm{CoAd}_G(G)\ell_1.$$

Proof. Suppose that the monomial representations $(U_{\ell_j,\mathfrak{h}_j}, \mathcal{H}_j)$ of G, $j \in \{1,2\}$, are unitarily isomorphic. Then we have the identity

$$\mathrm{tr}\ U^1_{\ell_1,\mathfrak{h}_1}(\phi) = \mathrm{tr}\ U^1_{\ell_2,\mathfrak{h}_2}(\phi)$$

for all functions $\phi \in \mathcal{D}(G)$. An application of the Kirillov character formula (Theorem 6.19) establishes the existence of measures β_j on the coadjoint orbits $\mathrm{CoAd}_G(G)\ell_j$ of G such that

$$\int_{\mathrm{CoAd}_G(G)\ell_1} \psi(\ell)d\beta_1(\ell) = \int_{\mathrm{CoAd}_G(G)\ell_2} \psi(\ell)d\beta_2(\ell)$$

holds for all functions $\psi \in \bar{\mathcal{F}}_{\mathfrak{g}}(\mathcal{D}(\mathfrak{g}))$. It follows $\beta_1 = \beta_2$, hence $\mathrm{CoAd}_G(G)\ell_1 = \mathrm{CoAd}_G(G)\ell_2$. In particular, $\ell_2 \in \mathrm{CoAd}_G(G)\ell_1$.

Conversely suppose that $\mathrm{CoAd}_G(G)\ell_1 = \mathrm{CoAd}_G(G)\ell_2$ holds. It follows

$$\mathrm{tr}\ U^1_{\ell_2,\mathfrak{h}_2}(\phi) = \rho.\mathrm{tr}\ U^1_{\ell_1,\mathfrak{h}_1}(\phi) \qquad (\rho > 0)$$

for all functions $\phi \in \mathcal{D}(G)$. In view of Theorems 6.12 and 3.10 we conclude $\rho = 1$ and the existence of a unitary isomorphy of $(U_{\ell_1,\mathfrak{h}_1}, \mathcal{H}_1)$ onto $(U_{\ell_2,\mathfrak{h}_2}, \mathcal{H}_2)$. —

Let us summarize the preceding results by stating *the full Kirillov theorem.*

6.21 Theorem (Kirillov). Let G be a simply connected, nilpotent, real Lie group with Lie algebra $\mathfrak{g}$. If $\ell \in \mathfrak{g}^*$ there exists a real polarization $\mathfrak{h}$ of $\mathfrak{g}$ for ℓ such that the monomial representation $(U_{\ell,\mathfrak{h}}, \mathcal{H})$ of G is topologically irreducible and of trace class. If $\ell \in \mathfrak{g}^*$ denotes a $\mathbb{R}$-linear form on $\mathfrak{g}$ that belongs to the coadjoint orbit $\mathrm{CoAd}_G(G)\ell$ and $\mathfrak{h}'$ a real polarization of $\mathfrak{g}$ for ℓ', then the monomial representations $(U_{\ell,\mathfrak{h}},\mathcal{H})$ and $(U_{\ell',\mathfrak{h}'},\mathcal{H}')$ of G are unitarily isomorphic. Conversely, if $\mathfrak{h}$, $\mathfrak{h}'$ denote real polarizations of $\mathfrak{g}$ for $\ell \in \mathfrak{g}^*$ and $\ell' \in \mathfrak{g}^*$, respectively, such that the monomial representation $(U_{\ell,\mathfrak{h}},\mathcal{H})$ and $(U_{\ell',\mathfrak{h}'},\mathcal{H}')$ of G are unitarily isomorphic, then ℓ and ℓ' belong to the same coadjoint orbit of G in $\mathfrak{g}^*$. Finally, for each topologically irreducible, continuous, unitary linear representation $(U_1,\mathcal{H}_1)$ of G in the complex Hilbert space $\mathcal{H}_1$ there exists a unique coadjoint orbit $\mathcal{O}$ of G in $\mathfrak{g}^*$ such that for any linear form $\ell \in \mathcal{O}$ and each real polarization $\mathfrak{h}$ of $\mathfrak{g}$ for ℓ the representations $(U_1,\mathcal{H}_1)$ and $(U_{\ell,\mathfrak{h}},\mathcal{H})$ of G are unitarily isomorphic.

The bijection of the space $\mathfrak{g}^*/G$ of coadjoint orbits of G in $\mathfrak{g}^*$ onto the unitary dual $\hat{G}$ of G (cf. 1.4) given by the preceding theorem is called the *Kirillov correspondence* of G. It furnishes a parametrization of $\hat{G}$ by means of the coadjoint orbit space.

References

Guichardet, A. : Théorie de Mackey et méthode des orbites selon M. Duflo. Expo. Math. 3 (1985), 303-346.

Kirillov, A.A. : Unitary representations of nilpotent Lie groups. Uspekhi Mat. Nauk. 17 (1962), 57-110. Russian Math. Surveys 17 (1962), 53-104.

Pukanszky, L. : Leçons sur les représentations des groupes. Monographies de la Soc. math. de France, Vol. 2. Dunod, Paris 1967.

Wallach, N.R. : Symplectic geometry and Fourier analysis. Lie groups: History, frontiers and applications, Vol. V. Math. Sci. Press, Brookline, Massachussets 1977.

7 The real Heisenberg nilpotent Lie group (part II)

7.1 Let us apply the Kirillov coadjoint orbit theory as outlined in the preceding section to describe completely the harmonic analysis of the real Heisenberg nilpotent Lie group $\tilde{A}(\mathbb{R})$ in a geometric way. Recall from Section 5.5 that the three-dimensional real Heisenberg nilpotent Lie algebra $\mathfrak{n}$ is formed by the nilpotent matrices

$$\begin{pmatrix} 0 & a & c \\ 0 & 0 & b \\ 0 & 0 & 0 \end{pmatrix}$$

with real entries a,b,c. Let {X,Y,T} denote the canonical basis of $\mathfrak{n}$ such that the Heisenberg canonical commutation relations

$$[X,T] = [Y,T] = 0, \ [X,Y] = T$$

hold, then $\mathfrak{c} = \mathbb{R}.T$ is the one-dimensional centre of $\mathfrak{n}$ and the decomposition

$$\mathfrak{n} = W \oplus \mathfrak{c}$$

holds where $W = \mathbb{R}.X \oplus \mathbb{R}.Y$ is a two-dimensional vector subspace of $\mathfrak{n}$. The bracket operation of $\mathfrak{n}$ induces on W the symplectic form

$$(pX + qY, \ p'X + q'Y) \rightsquigarrow B(\begin{bmatrix} p \\ q \end{bmatrix}, \begin{bmatrix} p' \\ q' \end{bmatrix}) = \det \begin{pmatrix} p & p' \\ q & q' \end{pmatrix}$$

and {X,Y} is the standard symplectic basis of the symplectic vector space (W;B). The three-dimensional real Heisenberg nilpotent Lie group $\tilde{A}(\mathbb{R})$ in its dual pairing presentation is the group of all unipotent matrices

$$\begin{pmatrix} 1 & x & z \\ 0 & 1 & y \\ 0 & 0 & 1 \end{pmatrix} = (x,y,z)$$

with real entries x,y,z. $\tilde{A}(\mathbb{R})$ is up to isomorphy the unique, simply connected real Lie group such that

$$\mathfrak{n} = \mathrm{Lie}(\tilde{A}(\mathbb{R}))$$

holds. The exponential mapping $\exp_{\tilde{A}(\mathbb{R})}: \mathfrak{n} \to \tilde{A}(\mathbb{R})$ forms a global diffeomorphism which carries the centre $\mathfrak{c} = \mathbb{R}.T$ of $\mathfrak{n}$ onto the centre $\tilde{C} = \{(0,0,z) \mid z \in \mathbb{R}\}$ of $\tilde{A}(\mathbb{R})$. For the cross-section W to $\mathfrak{c}$ in $\mathfrak{n}$, an easy calculation shows the identity

$$\exp_{\tilde{A}(\mathbb{R})} W = \{(x,y,\tfrac{1}{2}xy) \mid x \in \mathbb{R},\ y \in \mathbb{R}\}.$$

This subset of $\tilde{A}(\mathbb{R})$ is called to be the *isotropic cross-section to* $\tilde{C}$ *in* $\tilde{A}(\mathbb{R})$. Apart from the isotropic cross-section which bridges the gap between the dual pairing presentation and the basic presentation of the real Heisenberg nilpotent Lie group $\tilde{A}(\mathbb{R})$ there is another important cross-section to the centre $\tilde{C}$ in $\tilde{A}(\mathbb{R})$, to wit, the *polarized cross-section*

$$\{(x,y,0) \mid x \in \mathbb{R},\ y \in \mathbb{R}\}.$$

Since this subset of $\tilde{A}(\mathbb{R})$ can be identified with the two-dimensional vector space $\mathbb{R} \oplus \mathbb{R}$ we may assume that the polarized cross-section to $\tilde{C}$ in $\tilde{A}(\mathbb{R})$ carries the two-dimensional Lebesgue measure $\mu \otimes \mu$. From the geometric point of view, $\tilde{A}(\mathbb{R})$ may be considered as a *differential principal fibre bundle* over the two-dimensional polarized resp. isotropic cross-section with structure group isomorphic to the one-dimensional centre $\tilde{C}$.

Let χ be a continuous unitary character of $\tilde{C}$ and $L^2(\tilde{A}(\mathbb{R})/\tilde{C},\chi;\mu \otimes \mu)$ the complex Hilbert space on which the monomial representation $\mathrm{Ind}_{\tilde{C}}^{\tilde{A}(\mathbb{R})}(\chi.\mathrm{id}_{\mathbb{C}},\mathbb{C})$ of $\tilde{A}(\mathbb{R})$ acts. Then we have the ascending filtration

$$\mathcal{K}(\tilde{A}(\mathbb{R})/\tilde{C}) \longrightarrow \mathcal{S}(\tilde{A}(\mathbb{R})/\tilde{C}) \longrightarrow L^2(\tilde{A}(\mathbb{R})/\tilde{C},\chi;\ \mu \otimes \mu)$$

of everywhere dense vector subspaces, where $\mathcal{K}(\tilde{A}(\mathbb{R})/\tilde{C})$ is isomorphic to the complex vector space $\mathcal{K}(\mathbb{R} \oplus \mathbb{R})$ of all continuous complex-valued functions on $\mathbb{R} \oplus \mathbb{R}$ with compact support, and $\mathcal{S}(\tilde{A}(\mathbb{R})/\tilde{C})$ is isomorphic to the Schwartz space $\mathcal{S}(\mathbb{R} \oplus \mathbb{R})$ of all infinitely differentiable complex-valued functions on $\mathbb{R} \oplus \mathbb{R}$ such that all their derivatives tend to zero at infinity, faster than any polynomial (cf. Section 5.8). Summarizing we get the following commutative diagram

$$\begin{array}{ccccc}
\mathcal{K}(W) & \hookrightarrow & \mathcal{S}(W) & \hookrightarrow & L^2(W;\mu \otimes \mu) \\
\downarrow & & \downarrow & & \downarrow \\
\mathcal{K}(\tilde{A}(\mathbb{R})/\tilde{C}) & \hookrightarrow & \mathcal{S}(\tilde{A}(\mathbb{R})/\tilde{C}) & \hookrightarrow & L^2(\tilde{A}(\mathbb{R})/\tilde{C},\chi;\ \mu \otimes \mu) \\
\downarrow & & \downarrow & & \downarrow \\
\mathcal{K}(\mathbb{R} \oplus \mathbb{R}) & \hookrightarrow & \mathcal{S}(\mathbb{R} \oplus \mathbb{R}) & \hookrightarrow & L^2(\mathbb{R} \oplus \mathbb{R};\mu \otimes \mu)
\end{array}$$

where the vertical arrows indicate isomorphisms.

The *convolution structure* of $L^2(\tilde{A}(\mathbb{R})/\tilde{C},\chi_1;\ \mu \otimes \mu)$ will be obtained by an extension of the convolution product

$$f * g(x',y') = \iint_{\mathbb{R} \oplus \mathbb{R}} f(x,y)g(x'-x,y'-y)\chi_1((x'-x)y))d\mu(x)d\mu(y)$$

of the complex algebra $\mathcal{S}(\tilde{A}(\mathbb{R})/\tilde{C})$. As in Section 5.8, $\chi_1 : \tilde{C} \to \mathbb{T}$ denotes the basic character of $\tilde{C}$, and the complex-valued functions f and g are elements of $\mathcal{S}(\tilde{A}(\mathbb{R})/\tilde{C})$. In this way, $L^2(\tilde{A}(\mathbb{R})/\tilde{C},\chi_1;\ \mu \otimes \mu)$ becomes a *complex convolution algebra*.

Similarly, the convolution structure of $L^2(W)$ will be obtained by an extension of the *fibred* convolution product

$$f * g(p',q') = \iint_W f(p,q)g(p'-p,q'-q)c_1((p',q'),(p,q))d\mu(p)d\mu(q)$$

where c_1 denotes the 2-cocycle of W in $\mathbb{T}$ introduced in Remark 2 of Section 1.2, and f,g are elements of $\mathcal{S}(W)$.

7.2 Returning to the three-dimensional real Heisenberg nilpotent Lie group $\tilde{A}(\mathbb{R})$ in its dual pairing representation, we get for all elements $(x,y,z) \in \tilde{A}(\mathbb{R})$ and $(x',y',z') \in \tilde{A}(\mathbb{R})$ the identity (cf. Section 2.11)

$$(\mathrm{Int}_{\tilde{A}(\mathbb{R})}(x,y,z))(x',y',z') = (x',y',-yx' + xy' + z').$$

Thus the automorphism of $\mathfrak{n}$ (cf. Section 4.11)

$$\mathrm{Ad}_{\tilde{A}(\mathbb{R})}(x,y,z) : \mathfrak{n} \to \mathfrak{n}$$

admits the matrix

$$\begin{pmatrix} 1 & 0 & 0 \\ 0 & 1 & 0 \\ -y & x & 1 \end{pmatrix} \in \underline{\underline{SL}}(3,\mathbb{R})$$

with respect to the canonical basis $\{X,Y,T\}$ of $\mathfrak{n}$. It follows that the contragredient automorphism

$$\mathrm{CoAd}_{\tilde{A}(\mathbb{R})}(x,y,z) : \mathfrak{n}^* \to \mathfrak{n}^*$$

of the dual vector space $\mathfrak{n}^*$ admits the unipotent matrix

$$\begin{pmatrix} 1 & 0 & y \\ 0 & 1 & -x \\ 0 & 0 & 1 \end{pmatrix}$$

with respect to the dual basis $\{X^*,Y^*,T^*\}$ of $\mathfrak{n}^*$. If the linear form $\ell \in \mathfrak{n}^*$ has the coordinates $(\rho,\kappa,\lambda) \in \mathbb{R}^3$ with respect to the basis $\{X^*,Y^*,T^*\}$ and its image under $\mathrm{CoAd}_{\tilde{A}(\mathbb{R})}(x,y,z)$ is the linear form $\ell' = \rho'X^*+\kappa'Y^*+\lambda'T^* \in \mathfrak{n}^*$, then the identities

$$\begin{cases} \rho' = \rho + y\lambda \\ \kappa' = \kappa - x\lambda \\ \lambda' = \lambda \end{cases}$$

follow. Therefore the coadjoint orbits of $\tilde{A}(\mathbb{R})$ in $\mathfrak{n}^*$ fall into two classes, namely

(I) The single points $\{(\rho,\kappa,0) \mid \rho \in \mathbb{R}, \kappa \in \mathbb{R}\}$ located in the plane $\lambda = 0$ spanned by the linear forms $\{X^*,Y^*\}$,

and

(II) The affine planes $\{(\mathbb{R}X^* + \mathbb{R}Y^* + \lambda T^*) \mid \lambda \in \mathbb{R}, \lambda \neq 0\}$ parallel to the homogeneous plane $\lambda = 0$ carrying the single point coadjoint orbits.

The following figure points out the two types of coadjoint orbits of $\tilde{A}(\mathbb{R})$ in $\mathfrak{n}^*$.

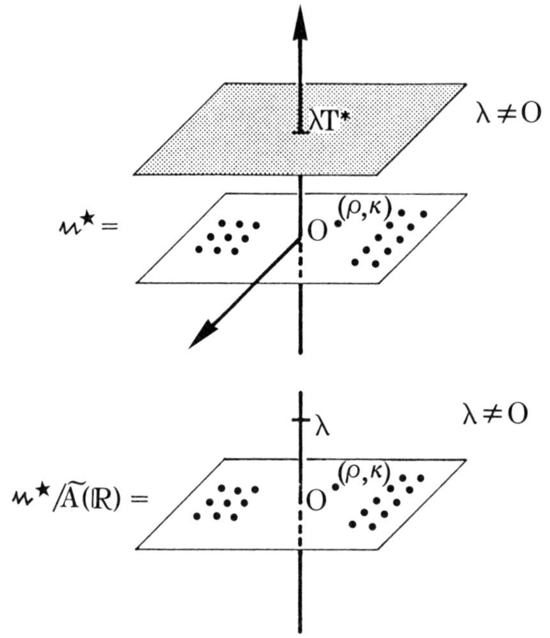

The single point coadjoint orbits of type (I) are called to be the *degenerate* orbits of $\tilde{A}(\mathbb{R})$ in $\mathfrak{n}^*$. The flat coadjoint orbits of type (II) are called to be the *non-degenerate* orbits of $\tilde{A}(\mathbb{R})$ in $\mathfrak{n}^*$. It follows that the degenerate coadjoint orbits of $\tilde{A}(\mathbb{R})$ in $\mathfrak{n}^*$ are located on the annihilator $\mathfrak{c}^0$ of the centre $\mathfrak{c}$ of $\mathfrak{n}^*$ whereas the non-degenerate coadjoint orbits of $\tilde{A}(\mathbb{R})$ in $\mathfrak{n}^*$ are the affine planes $\mathfrak{c}^0 + \lambda T^*$ $(\lambda \neq 0)$. The homogeneous plane $\mathfrak{c}^0$ in $\mathfrak{n}^*$ is the *tangent space* of the non-degenerate coadjoint orbits of $\tilde{A}(\mathbb{R})$ in $\mathfrak{n}^*$ and isomorphic to the dual W* of the cross-section W to $\mathfrak{c}$ in $\mathfrak{n}$. Since W is a two-dimensional symplectic vector space, it is *self-dual* by virtue of its symplectic form B. Therefore the non-degenerate coadjoint orbits of $\tilde{A}(\mathbb{R})$ in $\mathfrak{n}^*$ are the two-dimensional, affine, symplectic vector spaces $W^* + \lambda T^*$ $(\lambda \neq 0)$ carrying the natural symplectic form

$$(\rho X^* + \kappa Y^* + \lambda T^*,\ \rho' X^* + \kappa' Y^* + \lambda T^*) \rightsquigarrow \lambda B\left(\begin{bmatrix}\rho\\ \kappa\end{bmatrix}, \begin{bmatrix}\rho'\\ \kappa'\end{bmatrix}\right).$$

According to Theorem 6.21 there is a bijection between the coadjoint orbits of $\tilde{A}(\mathbb{R})$ in $\mathfrak{n}^*$ and the unitary dual $\tilde{A}(\mathbb{R})^\wedge$ of $\tilde{A}(\mathbb{R})$. Consistently with the rough classification outlined in Section 4.6 and the classification by the Mackey machinery based on the semi-direct product decompositions of $\tilde{A}(\mathbb{R})$

(cf. Theorem 5.7), the Kirillov correspondence yields two types of isomorphy clsses of topologically irreducible, continuous, unitary, linear representations of $\tilde{A}(\mathbb{R})$. In order to make the Kirillov correspondence precise we have to determine the polarizations of $\mathfrak{n}$ for the linear forms $\ell \in \mathfrak{n}^*$.

7.3 Let us consider the three-dimensional real Heisenberg nilpotent Lie group $\tilde{A}(\mathbb{R})$ in its basic presentation and let $\ell = \rho X^* + \kappa Y^* + \lambda T^* \in \mathfrak{n}^*$ be a given linear form on its Lie algebra $\mathfrak{n}$.

(I) If $\ell \in \mathfrak{c}^0$, i.e., $\lambda = 0$, then $\mathfrak{h} = \mathfrak{n}$ is the unique polarization of $\mathfrak{n}$ for ℓ. Indeed, we have $[\mathfrak{h},\mathfrak{h}] = \mathfrak{c}$ and therefore (cf. Section 6.4)

$$B_\ell | (\mathfrak{h} \times \mathfrak{h}) = 0.$$

An application of Theorem 6.21 shows that the isomorphy classes of topologically irreducible, continuous, unitary linear representations of $\tilde{A}(\mathbb{R})$ associated with the *degenerate* coadjoint orbits $(\rho,\kappa,0) \in \mathfrak{c}^0$ of $\tilde{A}(\mathbb{R})$ in $\mathfrak{n}^*$ under the Kirillov correspondence have the *degenerate* linear representations $(U_{(\rho,\kappa)},\mathbb{C})$ of $\tilde{A}(\mathbb{R})$ with

$$U_{(\rho,\kappa)} : \tilde{A}(\mathbb{R}) \ni (x,y,z) \rightsquigarrow e^{2\pi i(\rho x + \kappa y)} \in \mathbb{T} \quad ((\rho,\kappa) \in \mathbb{R} \times \mathbb{R})$$

as representatives (cf. Section 5.5).

(II) If $\ell \notin \mathfrak{c}^0$, i.e., $\lambda \neq 0$, any maximal isotropic vector subspace of $\mathfrak{n}$ relative to B_ℓ is a Lie subalgebra of $\mathfrak{n}$, hence a polarization $\mathfrak{h}$ of $\mathfrak{n}$ for ℓ which includes $\mathfrak{c}$. Let W_0 denote a cross-section to $\mathfrak{c}$ in $\mathfrak{h}$. If $pX + qY$ and $p'X + q'Y$ are any two vectors of W_0 then we have

$$B_\ell(pX + qY, p'X + q'Y) = \lambda B\left(\begin{bmatrix} p \\ q \end{bmatrix}, \begin{bmatrix} p' \\ q' \end{bmatrix}\right) = 0.$$

It follows $\dim_{\mathbb{R}} W_0 = 1$. Consequently, every polarization of $\mathfrak{n}$ for ℓ takes the form

$$\mathfrak{h} = W_0 \oplus \mathfrak{c}$$

where W_0 is a *one-dimensional* vector subspace of W. In order to apply Theorem 6.21 we may choose $W_0 = \mathbb{R}.Y$. Using the notations of Section 5.5 we then have $\mathfrak{h} = \mathfrak{n}_0$, $H = N_0$, and the isomorphy classes of topologically irreducible, continuous, unitary, linear representation of $\tilde{A}(\mathbb{R})$ associated with the *non-degenerate* coadjoint orbits $\mathfrak{c}^0 + \lambda T^*$ $(\lambda \neq 0)$ of $\tilde{A}(\mathbb{R})$ in $\mathfrak{n}^*$ under the Kirillov correspondence have the *non-degenerate* linear representations $(U_\lambda, L^2(\mathbb{R}))$ as representatives by which $\tilde{A}(\mathbb{R})$ acts on the functions $\psi \in \mathcal{S}(\mathbb{R})$ according to

$$U_\lambda(x,y,z)\psi(t) = e^{2\pi i\lambda(z-yt+\frac{1}{2}xy)}\psi(t-x) \qquad (t \in \mathbb{R})$$

(cf. Section 5.9). Consider the central character

$$\chi_\ell : \tilde{C} \ni (0,0,z) \rightsquigarrow e^{2\pi i\lambda z} \in \mathbb{T} \quad (\lambda \neq 0)$$

of $(U_\lambda, L^2(\mathbb{R}))$. Since the mapping $\mathcal{S}(\tilde{A}(\mathbb{R})/\tilde{C}) \ni \phi \rightsquigarrow U_\lambda^1(\phi) \in \mathcal{L}_1(L^2(\mathbb{R}))$ extends by the Plancherel theorem to a *unitary isomorphism*

$$U_\lambda^2 : L^2(\tilde{A}(\mathbb{R})/\tilde{C}, \chi_\ell; \mu \otimes \mu) \rightarrow \mathcal{L}_2(L^2(\mathbb{R}))$$

of the representation space $L^2(\tilde{A}(\mathbb{R})/\tilde{C}, \chi_\ell; \mu \otimes \mu)$ of $\mathrm{Ind}_{\tilde{C}}^{\tilde{A}(\mathbb{R})}(\chi_\ell . \mathrm{id}_{\mathbb{C}}, \mathbb{C})$ onto the complex Hilbert space $\mathcal{L}_2(L^2(\mathbb{R}))$ of all Hilbert-Schmidt operators on $L^2(\mathbb{R})$, we established in view of Section 3.9 and Theorem 6.19 the following result.

7.4 Theorem. (Full Stone-von Neumann-Mackey-Kirillov theorem). For all values of the real parameter $\lambda \neq 0$, the non-degenerate linear representation $(U_\lambda, L^2(\mathbb{R}))$ of $\tilde{A}(\mathbb{R})$ is up to isomorphy the unique topologically irreducible, continuous, unitary, linear representation of $\tilde{A}(\mathbb{R})$ with central character χ_ℓ, and this representation of $\tilde{A}(\mathbb{R})$ is square integrable modulo its centre $\tilde{C}$. For all test functions $\phi \in \mathcal{D}(\tilde{A}(\mathbb{R}))$ the character formula

$$\mathrm{tr}\, U_\lambda^2(\phi) = \frac{1}{\lambda}\int_{\mathbb{R}} \phi(0,0,z)e^{2\pi i\lambda z}\, d\mu(z) \quad (\lambda \neq 0)$$

holds.

The preceding theorem shows that it is sufficient to look at the *central* characters of the topologically irreducible, continuous, unitary, linear representations of $\tilde{A}(\mathbb{R})$ for checking the isomorphy. Thus the *centre* $\tilde{C}$ of $\tilde{A}(\mathbb{R})$ plays a "central" role in the harmonic analysis of the real Heisenberg nilpotent Lie group $\tilde{A}(\mathbb{R})$. It can be proved that the square integrability modulo $\tilde{C}$ of the prototypes $(U_\lambda, L^2(\mathbb{R}))$ among the non-degenerate linear representations of $\tilde{A}(\mathbb{R})$ is equivalent to the *flatness* of the non-degenerate coadjoint orbits $\mathfrak{c}^0 + \lambda T^* \in \mathfrak{n}^*/\tilde{A}(\mathbb{R})$ $(\lambda \neq 0)$ associated by the Kirillov correspondence. If the real parameter λ is normalized by setting $\lambda = 1$, the topologically irreducible, continuous, unitary, linear representation $(U, L^2(\mathbb{R}))$ by which $\tilde{A}(\mathbb{R})$ acts on the functions $\psi \in \mathcal{S}(\mathbb{R})$ according to the prescription

$$U(x,y,z)\psi(t) = e^{2\pi i(z+yt+\frac{1}{2}xy)}\psi(t-x) \qquad (t \in \mathbb{R})$$

for all elements $(x,y,z) \in \tilde{A}(\mathbb{R})$, is unitarily isomorphic to $(U_1, L^2(\mathbb{R}))$. Thus $(U, L^2(\mathbb{R}))$ which will be called to be the linear *Schrödinger representation of* $\tilde{A}(\mathbb{R})$ is up to dilation and isomorphy the unique non-degenerate linear representation of $\tilde{A}(\mathbb{R})$. The coadjoint orbit $W^* + T^*$ associated with $(U, L^2(\mathbb{R}))$ by the Kirillov correspondence carries the symplectic form B which occurs in the basic presentation of $\tilde{A}(\mathbb{R})$.

7.5 Let us exploit the fundamental Theorem 7.4 supra. According to Section 7.1 the cross-section W to $\mathfrak{c}$ in $\mathfrak{n}$ is a two-dimensional symplectic vector space $(W;B)$. Form the isometry group of $(W;B)$, i.e., the real symplectic group

$$\underline{\underline{Sp}}(W;B) = \underline{\underline{Sp}}(1,\mathbb{R}) = \{\sigma \in \underline{\underline{GL}}(2,\mathbb{R}) \mid B(\sigma\begin{bmatrix}p\\q\end{bmatrix}, \sigma\begin{bmatrix}p'\\q'\end{bmatrix}) = B(\begin{bmatrix}p\\q\end{bmatrix}, \begin{bmatrix}p'\\q'\end{bmatrix}) \textit{ for all } (p,q) \in \mathbb{R} \oplus \mathbb{R},\ (p',q') \in \mathbb{R} \oplus \mathbb{R}\}.$$

Then we have

$$\underline{\underline{Sp}}(1,\mathbb{R}) = \underline{\underline{SL}}(2,\mathbb{R})$$

and the elements $\sigma \in \underline{\underline{Sp}}(1,\mathbb{R})$ extend to the automorphisms $\sigma \times \mathrm{id}_{\tilde{C}}$ of $\tilde{A}(\mathbb{R})$ leaving the centre $\tilde{C}$ of $\tilde{A}(\mathbb{R})$ pointwise fixed. It follows that the topologically irreducible, continuous, unitary linear representations $(U, L^2(\mathbb{R}))$ and $(U \circ (\sigma \times \mathrm{id}_{\mathbb{C}}), L^2(\mathbb{R}))$ of $\tilde{A}(\mathbb{R})$ are associated with the same coadjoint orbit $W^* + T^*$ in $\mathfrak{n}^*$ under the Kirillov correspondence and are therefore unitarily isomorphic. Consequently there is a unitary operator $T_\sigma \in \underline{U}(L^2(\mathbb{R}))$ such that

$$T_\sigma \circ U(\sigma(x,y),z) \circ T_\sigma^{-1} = U(x,y,z)$$

holds for all elements $(x,y,z) \in \tilde{A}(\mathbb{R})$. This identity will be called the *metaplectic* (or *covariance*) *formula* for the linear Schrödinger representation

$(U, L^2(\mathbb{R}))$ of $\tilde{A}(\mathbb{R})$. If, for instance, $j \in \underline{\underline{Sp}}(1,\mathbb{R})$ has the matrix

$$\begin{pmatrix} 0 & 1 \\ -1 & 0 \end{pmatrix}$$

with respect to the basis $\{X,Y\}$ of W, then j is called to be the *Weyl element* of $\underline{\underline{Sp}}(1,\mathbb{R})$ and we obtain the automorphism (cf. Section 5.9)

$$J = j \times id_{\tilde{C}} : (x,y,z) \rightsquigarrow (y,-x,z)$$

of $\tilde{A}(\mathbb{R})$ which leaves the centre $\tilde{C}$ pointwise fixed. Since $U \circ J = V_1'$, the unitary operator $T_j \in \underline{\underline{U}}(L^2(\mathbb{R}))$ associated with the Weyl element $j \in \underline{\underline{Sp}}(1,\mathbb{R})$ coincides with the Fourier cotransform $\bar{\mathcal{F}}_{\mathbb{R}}$.

The metaplectic formula suggests to consider the subgroup G of the unitary group $\underline{\underline{U}}(L^2(\mathbb{R}))$ consisting of all operators $T \in \underline{\underline{U}}(L^2(\mathbb{R}))$ such that for each element $(x,y,z) \in \tilde{A}(\mathbb{R})$ there is an element $(x',y',z) \in \tilde{A}(\mathbb{R})$ having the same projection $(0,0,z) \in \tilde{C}$ onto the centre $\tilde{C}$ such that

$$T \circ U(x',y',z) \circ T^{-1} = U(x,y,z)$$

holds. In order to investigate the dependence of $(x',y') \in \mathbb{R} \oplus \mathbb{R}$ on $(x,y) \in \mathbb{R} \oplus \mathbb{R}$ and $T \in G$ we equip $\underline{\underline{U}}(L^2(\mathbb{R}))$ with the strong operator topology, i.e., the finest topology such that the evaluation mapping

$$\underline{\underline{U}}(L^2(\mathbb{R})) \times L^2(\mathbb{R}) \ni (T,f) \rightsquigarrow T(f) \in L^2(\mathbb{R})$$

remains continuous and establish the following technical

7.6 Lemma. The injective continuous mapping

$$(x,y,0) \rightsquigarrow U(x,y,0)$$

is a homeomorphism which embeds the polarized cross-section to the centre $\tilde{C}$ in $\tilde{A}(\mathbb{R})$ into the unitary group $\underline{\underline{U}}(L^2(\mathbb{R}))$.

Proof. It is sufficient to show that the image of every neighbourhood of the origin of $\mathbb{R} \oplus \mathbb{R}$ is a neighbourhood of the identity element $id_{L^2(\mathbb{R})}$ of $\underline{\underline{U}}(L^2(\mathbb{R}))$. If $\psi \in \mathbb{K}(\mathbb{R})$ satisfies $\|\psi\| = 1$ and

$$\|U(x,y,0)\psi - \psi\| < \sqrt{2}$$

for a fixed point $(x,y) \in \mathbb{R} \oplus \mathbb{R}$, then we have

$$2 \operatorname{Re} \int_{\mathbb{R}} e^{2\pi i(yt+\frac{1}{2}xy)} \psi(t-x)\bar{\psi}(t)dt < 0.$$

Consequently there is a point $t \in \mathbb{R}$ such that $\psi(t-x)\bar{\psi}(t) \neq 0$. It follows $t \in \mathrm{Supp}(\psi)$ and $t-x = s \in \mathrm{Supp}(\psi)$. Consequently $|x| = |t-s|$ becomes small when $\mathrm{Supp}(\psi)$ is contained in a small neighbourhood of the origin of $\mathbb{R}$. The same argument applied to the Fourier cotransform $\bar{\mathcal{F}}_{\mathbb{R}}$ yields the result. —

Returning to the group of isomorphisms

$$G = \{T \in \underline{\underline{U}}(L^2(\mathbb{R})) \mid T \circ U(x',y',z) \circ T^{-1} = U(x,y,z),\ (x,y) \in \mathbb{R} \oplus \mathbb{R},$$
$$(x',y') \in \mathbb{R} \oplus \mathbb{R},\ z \in \mathbb{R}\},$$

an application of Lemma 7.6 shows that for all elements $T \in G$ there exists a homeomorphism

$$h(T) : \mathbb{R} \oplus \mathbb{R} \to \mathbb{R} \oplus \mathbb{R}$$

such that $(x',y') = h(T)(x,y)$. We will show that $h(T)$ is $\mathbb{R}$-linear, hence $h(T) \in \underline{\underline{GL}}(2,\mathbb{R})$ for all $T \in G$. More precisely we have the following result.

7.7 Theorem. The image of the mapping $h : G \to \underline{\underline{GL}}(2,\mathbb{R})$ is the symplectic group $\underline{\underline{Sp}}(1,\mathbb{R}) = \underline{\underline{SL}}(2,\mathbb{R})$ and the sequence

$$\{1\} \longrightarrow \underline{\underline{T}}.\mathrm{id}_{L^2(\mathbb{R})} \longrightarrow G \xrightarrow{h} \underline{\underline{Sp}}(1,\mathbb{R}) \longrightarrow \{1_{\underline{\underline{Sp}}(1,\mathbb{R})}\}$$

is exact.

Proof. For all elements $(x_1,y_1) \in \mathbb{R} \oplus \mathbb{R}$, $(x_2,y_2) \in \mathbb{R} \oplus \mathbb{R}$ and $T \in G$ we get the identities

$$T \circ U(h(T)(x_1,y_1)+h(T)(x_2,y_2),\ \tfrac{1}{2}\, B(h(T)\begin{bmatrix} x_1 \\ y_1 \end{bmatrix}, h(T)\begin{bmatrix} x_2 \\ y_2 \end{bmatrix})) \circ T^{-1}$$

$$= T \circ U(h(T)(x_1,y_1),0) \circ T^{-1} \circ T \circ U(h(T)(x_2,y_2),0) \circ T^{-1}$$

$$= U(x_1,y_1,0) \circ U(x_2,y_2,0)$$

and

$$T \circ U(h(T)(x_1+x_2, y_1+y_2), \tfrac{1}{2} B(\begin{bmatrix} x_1 \\ y_1 \end{bmatrix}, \begin{bmatrix} x_2 \\ y_2 \end{bmatrix})) \circ T^{-1}$$

$$= U(x_1+x_2, y_1+y_2, \tfrac{1}{2} B(\begin{bmatrix} x_1 \\ y_1 \end{bmatrix}, \begin{bmatrix} x_2 \\ y_2 \end{bmatrix}))$$

$$= U(x_1, y_1, 0) \circ U(x_2, y_2, 0).$$

Consequently

$$h(T)(x_1, y_1) + h(T)(x_2, y_2) = h(T)((x_1, y_1) + (x_2, y_2))$$

and

$$B(h(T) \begin{bmatrix} x_1 \\ y_1 \end{bmatrix}, h(T) \begin{bmatrix} x_2 \\ y_2 \end{bmatrix}) = B(\begin{bmatrix} x_1 \\ y_1 \end{bmatrix}, \begin{bmatrix} x_1 \\ y_2 \end{bmatrix}).$$

It follows $h(T) \in \underline{\underline{Sp}}(1,\mathbb{R})$ for all $T \in G$. The next step in the proof is to show that the mapping $h: G \to \underline{\underline{Sp}}(1,\mathbb{R})$ is surjective. Consider the mappings

$$u : \mathbb{R}^{\times} \ni a \rightsquigarrow \begin{pmatrix} a & 0 \\ 0 & a^{-1} \end{pmatrix} \in \underline{\underline{Sp}}(1,\mathbb{R}),$$

$$v : \mathbb{R} \ni b \rightsquigarrow \begin{pmatrix} 1 & 0 \\ b & 1 \end{pmatrix} \in \underline{\underline{Sp}}(1,\mathbb{R}).$$

Then u is a homomorphism of the multiplicative group $\mathbb{R}^{\times} = \mathbb{R} - \{0\}$ into $\underline{\underline{Sp}}(1,\mathbb{R})$. The Weyl element $j \in \underline{\underline{Sp}}(1,\mathbb{R})$ with $j^2 = u(-1)$ and $j^4 = u(1) = 1_{\underline{\underline{Sp}}(1,\mathbb{R})}$ acts on the subgroup $A = u(\mathbb{R}^{\times})$ of $\underline{\underline{Sp}}(1,\mathbb{R})$ by conjugation. Indeed we have

$$j.u(a).j^{-1} = u(a^{-1}) \qquad (a \in \mathbb{R}^{\times}).$$

The mapping v is a homomorphism of the additive group $\mathbb{R}$ into $\underline{\underline{Sp}}(1,\mathbb{R})$ and A acts on the subgroup $N = v(\mathbb{R})$ of $\underline{\underline{Sp}}(1,\mathbb{R})$ by conjugation:

$$u(a)v(b)u(a^{-1}) = v(ba^{-2}).$$

The subgroups N, A of $\underline{\underline{Sp}}(1,\mathbb{R})$ satisfy $NA = AN$ and yield the *Bruhat decomposition*

$$\underline{\underline{\mathrm{Sp}}}(1,\mathbb{R}) = NA \cup NAjN$$

of $\underline{\underline{\mathrm{Sp}}}(1,\mathbb{R})$. If $f \in \mathcal{S}(\mathbb{R})$ and $a \in \mathbb{R}^{\times}$ define the scaling operator $\alpha(a) \in G$ by the prescription

$$(a)f : \mathbb{R} \ni t \rightsquigarrow \frac{1}{\sqrt{|a|}}\, f\left(\frac{1}{a}t\right) \in \mathbb{C}$$

and if $b \in \mathbb{R}$ define the multiplication operator $\beta(b) \in G$ according to the rule

$$\beta(b)f : \mathbb{R} \ni t \rightsquigarrow e^{\pi i b t^2} f(t) \in \mathbb{C}.$$

Then we get the identities

$$\begin{cases} h \circ \alpha = u, \\ h \circ \beta = v, \\ h(\bar{\mathcal{F}}_{\mathbb{R}}) = j, \end{cases}$$

so that $h : G \to \underline{\underline{\mathrm{Sp}}}(1,\mathbb{R})$ forms by Lemma 7.6 a continuous surjective morphism of the subgroup G of $\underline{U}(L^2(\mathbb{R}))$ onto the locally compact topological group $\underline{\underline{\mathrm{Sp}}}(1,\mathbb{R})$. Obviously $\mathrm{Ker}(h)$ consists of the unitary automorphisms of $(U, L^2(\mathbb{R}))$, i.e., we have

$$\mathrm{Ker}(h) = \mathcal{R}_{\tilde{A}(\mathbb{R})}(U) \cap \underline{U}(L^2(\mathbb{R})).$$

The linear Schrödinger representation $(U, L^2(\mathbb{R}))$ of $\tilde{A}(\mathbb{R})$ is topologically irreducible by Theorem 5.6, and Corollary 2 of Theorem 1.6, thus

$$\mathrm{Ker}(h) = \mathbb{T}.\mathrm{id}_{L^2(\mathbb{R})}.$$

This completes the proof. —

7.8 As we noted above, we get for each $\sigma \in \underline{\underline{\mathrm{Sp}}}(1,\mathbb{R})$ an operator $T_{\sigma} \in \underline{U}(L^2(\mathbb{R}))$ which is defined up to multiples $\zeta \in \mathbb{T}$ and satisfies the metaplectic formula for the linear Schrödinger representation $(U, L^2(\mathbb{R}))$ of $\tilde{A}(\mathbb{R})$. The action

$$\underline{\underline{\mathrm{Sp}}}(1,\mathbb{R}) \ni \sigma \rightsquigarrow T_{\sigma} \in \underline{U}(L^2(\mathbb{R}))$$

defines a continuous, unitary, *projective*, linear representation of the symplectic group $\underline{\underline{Sp}}(1,\mathbb{R}) = \underline{\underline{SL}}(2,\mathbb{R})$ in the complex Hilbert space $L^2(\mathbb{R})$ (cf. Section 1.2, Remark 2). Let $c : \underline{\underline{Sp}}(1,\mathbb{R}) \times \underline{\underline{Sp}}(1,\mathbb{R}) \to \mathbb{T}$ denote the associated 2-cocycle on $\underline{\underline{Sp}}(1,\mathbb{R})$ and $\underline{\underline{Sp}}(1,\mathbb{R})_c$ the Mackey obstruction group. It can be proved that the restriction of the continuous, unitary, linear representation $(U_c, L^2(\mathbb{R}))$ of $Sp(1,\mathbb{R})_c$ in $L^2(\mathbb{R})$ onto the subgroup

$$\underline{\underline{Mp}}(1,\mathbb{R}) = \underline{\underline{Sp}}(1,\mathbb{R}) \times \{-1, +1\}$$

forms a true, continuous, unitary, linear representation of $\underline{\underline{Mp}}(1,\mathbb{R})$ in $L^2(\mathbb{R})$. The double covering $\underline{\underline{Mp}}(1,\mathbb{R})$ of $\underline{\underline{Sp}}(1,\mathbb{R})$ is called to be the real *metaplectic group*. Let

$$\underline{\underline{Mp}}(1,\mathbb{R}) \ni \tilde{\sigma} \leadsto \sigma \in \underline{\underline{Sp}}(1,\mathbb{R})$$

denote the covering homomorphism so that each fibre consists of two points. Then the continuous, unitary, projective, linear representation $\sigma \leadsto T_\sigma$ of $\underline{\underline{Sp}}(1,\mathbb{R})$ in $L^2(\mathbb{R})$ lifts to the Segal-Shale-Weil *metaplectic* (or linear *oscillator*) *representation*

$$\underline{\underline{Mp}}(1,\mathbb{R}) \ni \tilde{\sigma} \leadsto T_{\tilde{\sigma}} \in \underline{\underline{U}}(L^2(\mathbb{R}))$$

of $\underline{\underline{Mp}}(1,\mathbb{R})$ in $L^2(\mathbb{R})$, and the metaplectic (or covariance) formula for the linear Schrödinger representation $(U, L^2(\mathbb{R}))$ of $\tilde{A}(\mathbb{R})$ takes the form

$$T_{\tilde{\sigma}} \circ U(\sigma(x,y),z) \circ T_{\tilde{\sigma}}^{-1} = U(x,y,z)$$

for all $\sigma \in \underline{\underline{Sp}}(1,\mathbb{R})$ and all elements $(x,y,z) \in \tilde{A}(\mathbb{R})$. Again $\underline{\underline{Mp}}(1,\mathbb{R})$ is a real Lie group. Its Lie algebra coincides with the Lie algebra $\mathfrak{sp}(1,\mathbb{R})$ of $\underline{\underline{Sp}}(1,\mathbb{R})$ and can therefore be identified with the Lie algebra $\mathfrak{sl}(2,\mathbb{R})$ of $\underline{\underline{SL}}(2,\mathbb{R})$ consisting of all traceless 2×2 matrices

$$\begin{pmatrix} a & b \\ c & d \end{pmatrix} \qquad (a + d = 0)$$

with real entries a,b,c,d.

7.9 If $\{X,Y,T\}$ denotes the canonical basis of the three-dimensional real Heisenberg Lie algebra $\mathfrak{n}$ then $\{X,Y\}$ forms a symplectic basis of the two-dimensional real symplectic vector space $(W;B)$:

$$B(X,Y) = 1 \qquad B(Y,X) = -1.$$

Since the Weyl element $j \in \underline{\underline{Sp}}(W;B) = \underline{\underline{Sp}}(1,\mathbb{R})$ satisfies

$$j^2 = -u(1),$$

it therefore defines on W a *complex structure* by letting the numbers $(a+ib)\in\mathbb{C}$ act on the vectors $qX + pY \in W$ according to the prescription

$$(a + ib).(pX + qY) = a(pX + qY) - bj(pX + qY)$$

where $(a,b) \in \mathbb{R} \oplus \mathbb{R}$ and $(p,q) \in \mathbb{R} \oplus \mathbb{R}$. Thus we may identify the vector

$$pX + qY = \begin{pmatrix} 0 & p & 0 \\ 0 & 0 & q \\ 0 & 0 & 0 \end{pmatrix} \in W$$

with the complex number

$$w = p + iq \in \mathbb{C}.$$

The linear Schrödinger representation $(U,L^2(\mathbb{R}))$ will be *adapted* to the complex structure of the symplectic vector space $(W;B)$ by considering the unitarily isomorphic linear representation $(V,L^2(\mathbb{R}))$ of $\tilde{A}(\mathbb{R})$ by which $\tilde{A}(\mathbb{R})$ acts on the functions $\psi \in \mathcal{S}(\mathbb{R})$ according to the prescription

$$V(x,y,z)\psi(t) = e^{2\pi i(z+yt-\frac{1}{2}xy)}\psi(t-x) \quad (t \in \mathbb{R})$$

for all elements $(x,y,z) \in \tilde{A}(\mathbb{R})$.

7.10 Let us lift the topologically irreducible, continuous, unitary, linear representation $(V,L^2(\mathbb{R}))$ to the cross-section W to the centre $\mathfrak{c}$ in $\mathfrak{n}$. For all functions $\psi \in \mathcal{S}(\mathbb{R})$ the identities

$$\begin{cases} V(X)\psi = \lim\limits_{\substack{s\to 0\\ s\neq 0}} \frac{1}{s}\left(V(\exp_{\tilde{A}(\mathbb{R})} sX)(\psi)-\psi\right) = -\frac{d}{dt}\psi, \\ V(Y)\psi = \lim\limits_{\substack{s\to 0\\ s\neq 0}} \frac{1}{s}\left(V(\exp_{\tilde{A}(\mathbb{R})} sY)(\psi)-\psi\right) = 2\pi i t.\psi \end{cases}$$

hold. Introduce the vectors

$$Z^- = \frac{1}{2}(X + iY), \quad Z^+ = \frac{1}{2}(X - iY)$$

of W as displayed in the following figure:

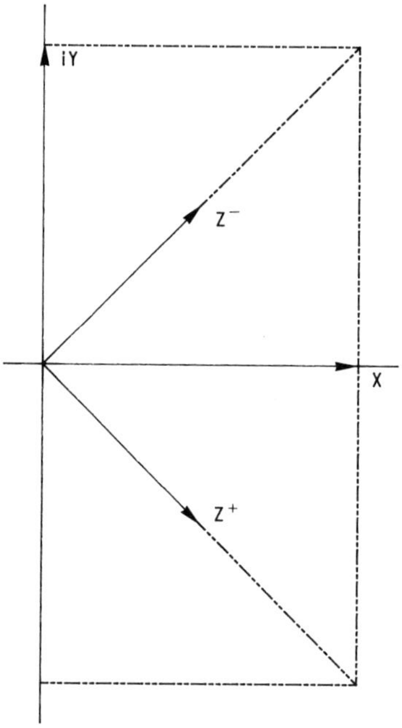

By evaluating the *differentiated* version of V at Z^- and Z^+, respectively, we get the linear maps

$$A^- := V(Z^-) = -\frac{1}{2}\left(\frac{d}{dt} + 2\pi t\right),$$

$$A^+ := V(Z^+) = -\frac{1}{2}\left(\frac{d}{dt} - 2\pi t\right),$$

of the vector subspace $\mathfrak{S}(\mathbb{R})$ of $L^2(\mathbb{R})$ into itself. It follows for the *adjoint* of A^-:

$$(A^-)^* = -\frac{1}{2}\left(-\frac{d}{dt} + 2\pi t\right)$$

$$= -A^+,$$

and moreover,

$$[A^+, A^-] = \mathrm{id}_{\mathfrak{S}(\mathbb{R})}.$$

Obviously the Gaussian

$$g_0 : \mathbb{R} \ni t \rightsquigarrow \sqrt[4]{2}\, e^{-\pi t^2} \in \mathbb{R}$$

belongs to the vector space $\mathcal{S}(\mathbb{R})$ and satisfies

$$\|g_0\| = 1, \qquad A^-(g_0) = 0.$$

Thus $g_0 \in \mathcal{S}(\mathbb{R})$ is a unit vector in $L^2(\mathbb{R})$ which will be *annihilated* by the linear operator $A^- : \mathcal{S}(\mathbb{R}) \to \mathcal{S}(\mathbb{R})$. For all integers $m \geq 1$ define inductively the functions

$$g_m = A^+(g_{m-1}) \in \mathcal{S}(\mathbb{R}).$$

Then we have the identity

$$A^-(g_m) = -\pi m\, g_{m-1} \qquad (m \geq 1).$$

Indeed, in the case $m = 1$ we get by the annihilation property

$$A^-(g_1) = A^- \circ A^+(g_0)$$

$$= (A^+ \circ A^- - [A^+,A^-])(g_0)$$

$$= -\pi g_0.$$

Suppose that the identity $A^-(g_{m-1}) = -\pi(m-1)g_{m-2}$ holds for $m \geq 2$. Then in a similar way

$$A^-(g_m) = A^- \circ A^+(g_{m-1})$$

$$= -\pi(m-1)g_{m-1} - \pi g_{m-1}$$

$$= -\pi m\, g_{m-1},$$

as desired.

It follows for all integers $k > l \geq 0$ by the annihilation property

$$\langle g_k|g_l\rangle = (-1)^k\langle g_0|(A^-)^k \circ (A^+)^l(g_0)\rangle$$

$$= (-1)^k\langle g_0|(A^-)^k(g_l)\rangle$$

$$= 0$$

and therefore also $\langle g_k|g_l\rangle = 0$ for $l > k \geq 0$. In the case $k = l = m \geq 0$, however, we get

$$\|g_m\|^2 = \pi^m m!$$

Normalizing the functions $(g_m)_{m \geqq 0}$ we obtain the sequence $(h_m)_{m \geqq 0}$ of functions

$$h_m = \frac{1}{\sqrt{\pi^m m!}} g_m = \frac{1}{\sqrt{\pi^m m!}} (A^+)^m(g_0) \in \mathcal{S}(\mathbb{R}) \qquad (m \geqq 0)$$

which form an *orthonormal family in* $L^2(\mathbb{R})$. The sequence $(h_m)_{m \geqq 0}$ satisfies the identities

$$A^-(h_m) = -\sqrt{\pi m}\, h_{m-1} \qquad (m \geqq 1),$$

$$A^+(h_m) = \sqrt{\pi(m+1)}\, h_{m+1} \qquad (m \geqq 0).$$

It follows for $m \geqq 0$

$$A^- \circ A^+(h_m) = -\pi(m+1)h_m,$$

$$A^+ \circ A^-(h_m) = -\pi m h_m.$$

The linear differential operator $A^+ \circ A^- = \frac{1}{4}\left(\frac{d^2}{dt^2} - 4\pi^2t^2 + 2\pi\right)$ is called to be the *number operator* for the family $(h_m)_{m \geqq 0}$. Moreover, we obtain

$$-(A^+ + A^-)(h_m) = \frac{d}{dt} h_m = \sqrt{\pi}\,(\sqrt{m}\, h_{m-1} - \sqrt{m+1}\, h_{m+1}) \qquad (m \geqq 1),$$

$$-(A^+ - A^-)(h_m) = 2\pi t.h_m = \sqrt{\pi}\,(\sqrt{m}\, h_{m-1} + \sqrt{m+1}\, h_{m+1}) \quad (m \geqq 1),$$

$$-\frac{1}{2}(A^+ \circ A^- + A^- \circ A^+)(h_m) = \frac{1}{4}\left(-\frac{d^2}{dt^2} + 4\pi^2t^2\right)h_m = \pi\left(m + \frac{1}{2}\right)h_m \quad (m \geqq 0).$$

In particular we conclude that the functions $(h_m)_{m \geqq 0}$ are *normalized eigenfunctions* of the linear differential operator of second order with constant coefficients

$$-2(A^+ \circ A^- + A^- \circ A^+) = \left(-\frac{d^2}{dt^2} + 4\pi^2t^2\right)$$

acting on the everywhere dense vector subspace $\mathcal{S}(\mathbb{R})$ of $L^2(\mathbb{R})$. This differential operator, the so-called *Hermite differential operator*, is essentially self-adjoint in the complex Hilbert space $L^2(\mathbb{R})$. Its closure, i.e., its minimal closed linear extension is a linear operator with pure point spectrum. The eigenvalues are $\{2\pi(2m+1) \mid m \in \mathbb{N}\}$ and all of these are simple. The associated sequence of orthonormal eigenfunctions $(h_m)_{m \geq 0}$ is formed by the normalized *Hermite functions* or *harmonic oscillator wave functions* in $\mathcal{S}(\mathbb{R})$. They are also known as the standard description for the modes of laser beams in square-law media and laser resonators (*Hermite-Gaussian eigenmodes*). In quantum mechanics, the linear differential operator $-\frac{1}{2}(A^+ \circ A^- + A^- \circ A^+)$ is called to be the *Schrödinger Hamiltonian* for the harmonic oscillator system. Its *eigenstates* are the harmonic oscillator wave functions $(h_m)_{m \geq 0}$ with equidistant *energies* $\{\pi(m + \frac{1}{2}) \mid m \in \mathbb{N}\}$ in natural units. The energy being quantized in units, the linear operators A^+ and A^- act as *creation* (or *raising*) and *annihilation* (or *lowering*) *operators* of energy quanta for the system. The first six harmonic oscillator wave functions $(h_m)_{0 \leq m \leq 5}$ are

$$\begin{aligned}
h_0(t) &= \sqrt[4]{2}\, e^{-\pi t^2} \\
h_1(t) &= 2\sqrt[4]{2}\,\sqrt{\pi}\, t\, e^{-\pi t^2} \\
h_2(t) &= \frac{1}{\sqrt[4]{2}}\,(4\pi t^2 - 1) e^{-\pi t^2} && (t \in \mathbb{R}) \\
h_3(t) &= \sqrt{\frac{2\pi}{3}}\,\sqrt[4]{2}\,(4\pi t^3 - 3t) e^{-\pi t^2} \\
h_4(t) &= \frac{\sqrt[4]{2}}{2\sqrt{6}}\,(16\pi^2 t^4 - 24\pi t^2 + 3) e^{-\pi t^2} \\
h_5(t) &= \sqrt{\frac{\pi}{30}}\,\sqrt[4]{2}\,(16\pi^2 t^5 - 40\pi t^3 + 15t) e^{-\pi t^2} \\
&\;\;\vdots
\end{aligned}$$

Starting with the *ground state* $h_0 \in \mathcal{S}(\mathbb{R})$ one can go on computing the eigenstates $(h_m)_{m \geq 1}$ by successive application of the creation operator A^+ and the normalizing procedure as long as patience will permit. We get $h_m(-t) = (-1)^m\, h_m(t)$ for $m \geq 0$ and all $t \in \mathbb{R}$ and furthermore

$$\frac{d}{dt} h_m(t) - 2\pi t h_m(t) = -2\sqrt{\pi}\,\sqrt{m+1}\; h_{m+1}(t).$$

Thus $(i)^m h_m$ and the Fourier cotransform $\bar{\mathcal{F}}_{\mathbb{R}} h_m$ obey the same recursion formula for $m \geqq 0$. In view of $\bar{\mathcal{F}}_{\mathbb{R}} h_0 = h_0 = g_0$ we get

$$\bar{\mathcal{F}}_{\mathbb{R}} h_m = (i)^m h_m \qquad (m \geqq 0)$$

and by taking complex conjugates,

$$\mathcal{F}_{\mathbb{R}} h_m = (-i)^m h_m \qquad (m \geqq 0).$$

Moreover

$$h_m(t) = p_m(t)\, e^{-\pi t^2} \qquad (t \in \mathbb{R})$$

where $(p_m)_{m \geqq 0}$ are the *Hermite polynomials* of precise degree $m \geqq 0$ normalized in such a way that they form an orthonormal family in the real Hilbert space $L^2(\mathbb{R}; e^{-2\pi t^2}\, d\mu(t))$. In the following figures the graphs of the first six oscillator wave functions $(h_m)_{0 \leqq m \leqq 5}$ are plotted.

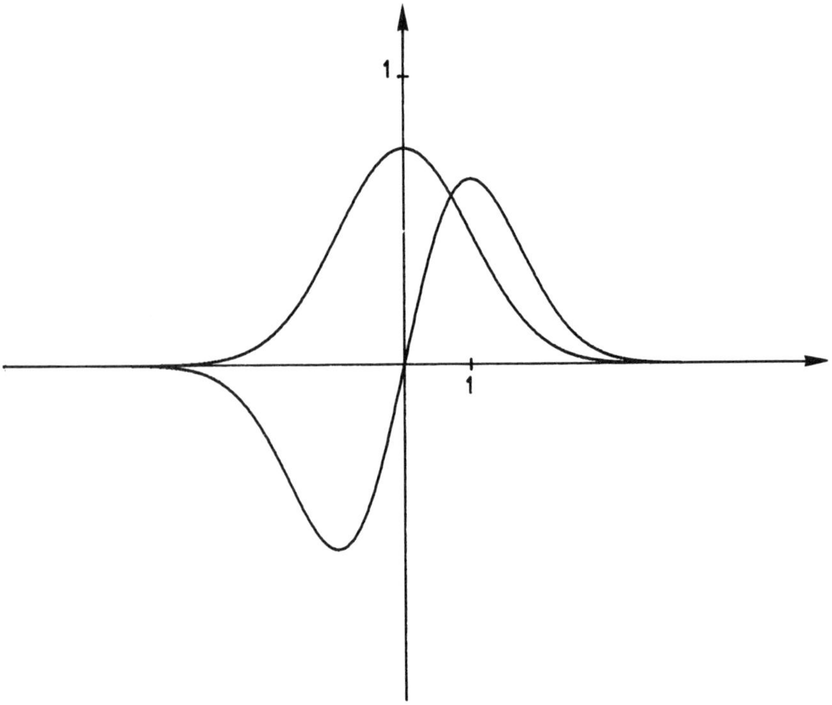

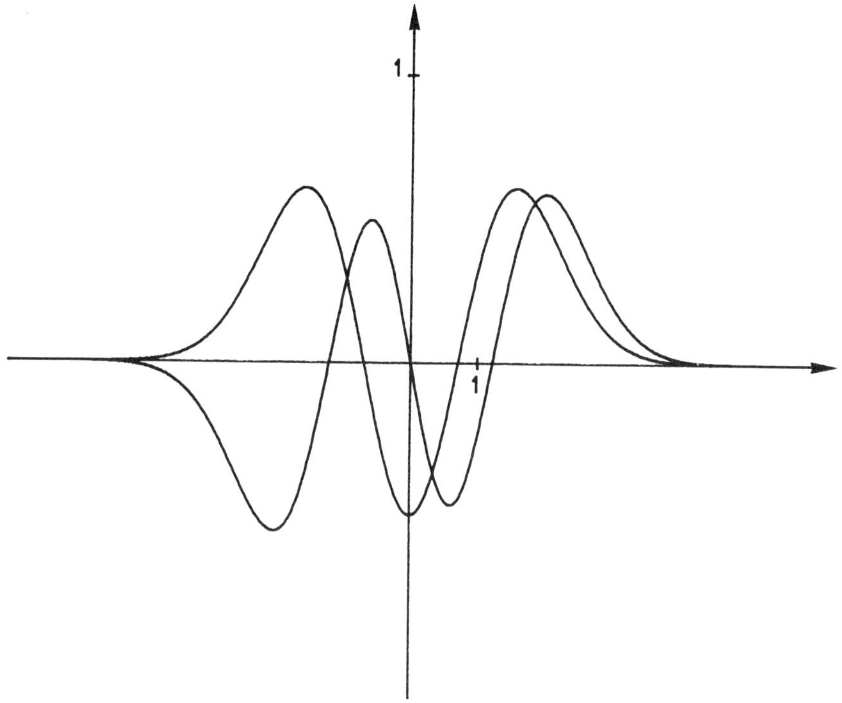
1
1

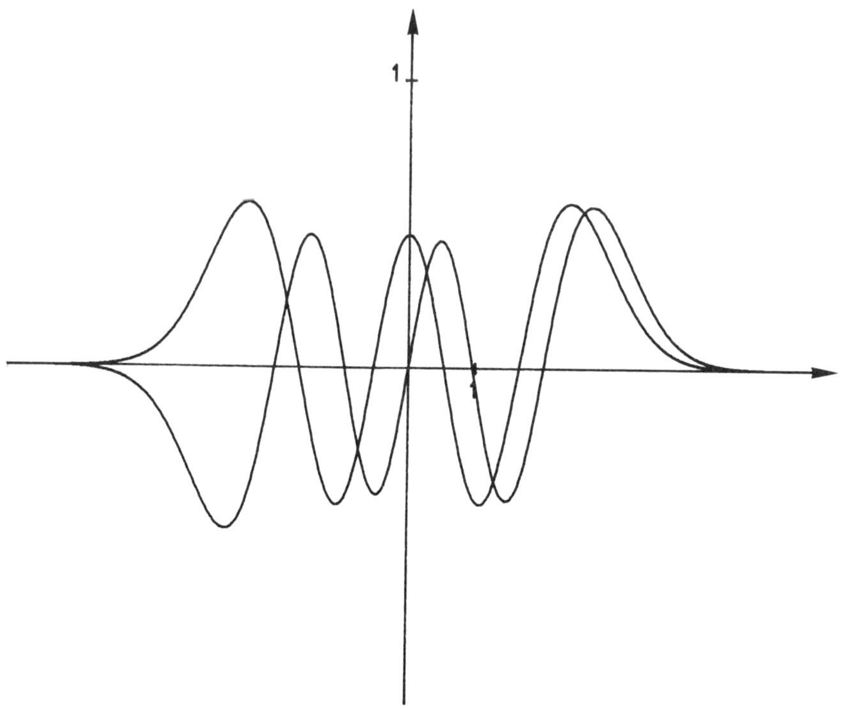
1

7.11 Theorem. The sequence $(h_m)_{m \geqq 0}$ of normalized Hermite functions in $\mathcal{S}(\mathbb{R})$ forms a Hilbert basis of $L^2(\mathbb{R})$, the elements of which are eigenfunctions of the Fourier transform $\mathcal{F}_{\mathbb{R}}$ and the Fourier cotransform $\bar{\mathcal{F}}_{\mathbb{R}}$ to the eigenvalues $(-i)^m$ and $(i)^m$, respectively.

Proof. Retain the above notations. For any vector $pX + qY \in W$ where $(p,q) \in \mathbb{R} \oplus \mathbb{R}$ and $w = p + iq \in \mathbb{C}$, define for all integers $k \geqq 0$, $m \geqq 0$ with $0 \leqq k \leqq m$ the complex numbers $c_{k,m}$ via the prescription

$$V(pX + qY)^m(g_0) = \sum_{0 \leqq k \leqq m} c_{k,m} g_k.$$

In view of $g_{-1} = A^-(g_0) = 0$ we get

$$\begin{aligned} V(pX + qY)^{m+1}(g_0) &= (wA^+ + \bar{w}A^-)\Big(\sum_{0 \leqq k \leqq m} c_{k,m} g_k\Big) \\ &= \sum_{0 \leqq k \leqq m} c_{k,m}(w g_{k+1} - \pi k \bar{w} g_{k-1}) \end{aligned}$$

and therefore

$$c_{k,m+1} = (w c_{k-1,m} - \pi(k+1)\bar{w} c_{k+1,m}) \qquad (1 \leqq k \leqq m-1).$$

Let $w = |w|e^{2\pi i\phi}$ be the representation of $w \neq 0$ in polar coordinates where $\phi \in [0,1[$ and

$$d_{k,m} = \Big(\frac{1}{\sqrt{|w|}} e^{-\pi i\phi}\Big)^{(m+k)} \Big(\frac{1}{\sqrt{\pi|w|}} e^{\pi i(\phi-1)}\Big)^{(m-k)} c_{m,k}.$$

Then the double recurrence

$$d_{k,m+1} = d_{k-1,m} + (k+1)d_{k+1,m} \qquad (1 \leqq k \leqq m-1)$$

pops up. For $1 \leqq k \leqq m-1$ set

$$b_{k,m} = d_{m-k,m}.$$

Then we get the three term recurrence

$$b_{k,m} = b_{k,m-1} + (m-k+1)b_{k-2,m-1} \qquad (2 \leqq k \leqq m-1).$$

If the starting value is

$$b_{0,0} = 0$$

and we define $b_{k,o} = 0$ for $k \geqq 1$ then we get

$$\begin{cases} b_{2l+1,m} = 0, & (0 \leqq l \leqq \frac{1}{2}(m-1)), \\ b_{2l,m} = \dfrac{m!}{2^l l!(m-2l)!} & (0 \leqq l \leqq \frac{1}{2}m). \end{cases}$$

Switching back we obtain

$$V(pX + qY)^m(h_o) = \sum_{0 \leqq l \leqq \frac{1}{2}m} (-\tfrac{1}{2}\pi\bar{w})^l w^{(m-l)} \frac{m!}{l!(m-2l)!} g_{m-2l}$$

and therefore

$$\begin{aligned} e^{V(pX+qY)}(h_o) &= \sum_{m \geqq 0} \frac{1}{m!} V(pX + qY)^m(h_o) \\ &= \sum_{k \geqq 0} (\frac{1}{k!} \sum_{l \geqq 0} \frac{1}{l!}(-\tfrac{1}{2}\pi|w|^2)^l) w^k g_k \\ &= e^{-\frac{1}{2}\pi|w|^2} e^{wA^+}(h_o). \end{aligned}$$

It follows that $V(\exp_{\tilde{A}(\mathbb{R})}(pX + qY))(g_o)$ belongs for all pairs $(p,q) \in \mathbb{R} \oplus \mathbb{R}$ to the closed vector space of $L^2(\mathbb{R})$ spanned by the sequence $(h_m)_{m \geqq 0}$. Since $\exp_{\tilde{A}(\mathbb{R})}W$ is the isotropic cross-section to $\tilde{C}$ in $\tilde{A}(\mathbb{R})$ (cf. Section 7.1) and the continuous, unitary, linear representation $(V, L^2(\mathbb{R}))$ of $\tilde{A}(\mathbb{R})$ is topologically irreducible, the sequence $(h_m)_{m \geqq 0}$ forms a *total* family in $L^2(\mathbb{R})$. Thus the theorem follows by the preceding calculations. —

Corollary. The sequence $(p_m)_{m \geqq 0}$ of Hermite polynomials of precise degree m forms a Hilbert basis of $L^2(\mathbb{R}; e^{-2\pi t^2} d\mu(t))$.

In view of the fact that the linear representation $(V, L^2(\mathbb{R}))$ of $\tilde{A}(\mathbb{R})$ is square integrable modulo $\tilde{C}$ (Theorem 7.4), it follows by Section 3.6 that the functions

$$W \ni w = p+iq \rightsquigarrow c_{V,h_m,h_n}(\exp_{\tilde{A}(\mathbb{R})}(pX+qY)) \in \mathbb{C} \qquad ((m,n) \in \mathbb{N} \times \mathbb{N})$$

form a Hilbert basis of the closed vector subspace $\exp_{\tilde{A}(\mathbb{R})}\mathcal{L}_V$ of $L^2(\tilde{A}(\mathbb{R})/\tilde{C}, \chi_1; \mu \otimes \mu)$. The closure $\mathbb{H}(W)$ of the sequence of functions

$$w \rightsquigarrow c_{V,h_o,h_n}(\exp_{\tilde{A}(\mathbb{R})}(pX + qY)) \qquad (n \geqq 0)$$

in $L^2(\tilde{A}(\mathbb{R})/\tilde{C},\chi_1; \mu \otimes \mu)$ forms a complex Hilbert space which is isomorphic to $L^2(\mathbb{R})$. For all integers $n \geqq 0$ we obtain

$$\begin{aligned} c_{V,h_o,h_n}(\exp_{\tilde{A}(\mathbb{R})}(pX+qY)) &= \langle e^{V(pX+qy)}(h_o)|h_n\rangle \\ &= e^{-\frac{1}{2}\pi|w|^2} \langle e^{wA^+}(h_o)|h_n\rangle \\ &= \frac{1}{n!} e^{-\frac{1}{2}\pi|w|^2} w^n\langle (A^+)^n(g_o)|h_n\rangle \\ &= \sqrt{\frac{\pi^n}{n!}} \; e^{-\frac{1}{2}\pi|w|^2} w^n. \end{aligned}$$

It follows that $\mathbb{H}(W)$ is the complex Hilbert space of all functions $f:W \to \mathbb{C}$ that are *holomorphic* on W and satisfy the *square integrability condition*

$$\int_W |f(w)|^2 e^{-\pi|w|^2} dwd\bar{w} < + \infty.$$

The restriction $\gamma_{\tilde{A}(\mathbb{R})/\tilde{C}}|\exp_{\tilde{A}(\mathbb{R})}\mathcal{L}_V$ gives rise to the topologically irreducible, continuous, unitary, linear representation $(\gamma_W, \mathbb{H}(W))$ of $\tilde{A}(\mathbb{R})$ by which $\tilde{A}(\mathbb{R})$ acts on the functions $f \in \mathbb{H}(W)$ according to the prescription

$$\gamma_W(w)f(w') = \gamma_{\tilde{A}(\mathbb{R})/\tilde{C}}(\exp_{\tilde{A}(\mathbb{R})}w,0)f(w') = c_1(w';w)f(w'-w) \qquad (w' \in W)$$

for all elements $w \in W$. As in Section 7.1 supra, c_1 denotes the 2-cocycle of W in $\mathbb{T}$ introduced in Remark 2 of Section 1.2. In view of the polarization identity for the exponent of c_1

$$\pi i B(w';w) = \pi(w.\bar{w}' - \tfrac{1}{2}|w'|^2 - \tfrac{1}{2}|w|^2 + \tfrac{1}{2}|w' - w|^2),$$

the action of $\tilde{A}(\mathbb{R})$ by γ_W on the functions $f \in \mathbb{H}(W)$ reads as follows:

$$\gamma_W(w)f(w') = e^{\pi(w.\bar{w}')} e^{-\frac{1}{2}\pi|w|^2} f(w'-w) \qquad (w' \in W)$$

for all $w \in W$. The linear representation $(\gamma_W, \mathcal{H}(W))$ of $\tilde{A}(\mathbb{R})$ is isomorphic to the linear Schrödinger representation $(U, L^2(\mathbb{R}))$ of $\tilde{A}(\mathbb{R})$; cf. Section 7.4. It is called to be the *linear Bargmann-Fock-Segal representation* of the real Heisenberg nilpotent Lie group $\tilde{A}(\mathbb{R})$. Of course, $(\gamma_W, \mathcal{H}(W))$ is isomorphic to the topologically irreducible, continuous, unitary, linear representation $(\delta_W, \mathcal{H}(W))$ of $\tilde{A}(\mathbb{R})$ by which $\tilde{A}(\mathbb{R})$ acts on the functions $f \in \mathcal{H}(W)$ according to the prescription

$$\delta_W(w)f(w') = e^{\pi(w.\bar{w}')} e^{-\frac{1}{2}\pi|w|^2} f(w'+w) \qquad (w' \in W)$$

for all $w \in W$. The linear representations $(\gamma_W, \mathcal{H}(W))$ and $(\delta_W, \mathcal{H}(W))$ of $\tilde{A}(\mathbb{R})$ are examples of *holomorphically induced*, topologically irreducible, continuous, unitary, linear representations.

7.12 Apart from the linear Schrödinger representation $(U, L^2(\mathbb{R}))$ and the linear Bargmann-Fock-Segal representation $(\gamma_W, \mathcal{H}(W))$ of $\tilde{A}(\mathbb{R})$ there is another important representative of the isomorphy class in $\tilde{A}(\mathbb{R})^{\wedge}_{(II)}$ which admits as its central character the basic character χ_1.

Consider the normal subgroup

$$D = \{(\xi,\eta,z) \in \tilde{A}(\mathbb{R}) \,|\, \xi \in \mathbb{Z},\ \eta \in \mathbb{Z}\}$$

of $\tilde{A}(\mathbb{R})$ with centre $\tilde{C}$ and the discrete normal subgroup

$$D_0 = \{(\xi,\eta,\zeta) \in \tilde{A}(\mathbb{R}) \,|\, \xi \in \mathbb{Z},\ \eta \in \mathbb{Z},\ \zeta \in \mathbb{Z}\}$$

of D. Thus we have the following inclusions of normal subgroups:

$$\begin{array}{ccccc} D_0 & \hookrightarrow & D & \hookrightarrow & \tilde{A}(\mathbb{R}) \\ \uparrow & & \uparrow & & \\ (D_0 \cap \tilde{C}) & \hookrightarrow & \tilde{C} & & \end{array}$$

Obviously D is the subgroup of $\tilde{A}(\mathbb{R})$ generated by $D_0 \cup \tilde{C}$ and the quotient group $D\backslash\tilde{A}(\mathbb{R}) = \tilde{A}(\mathbb{R})/D$ is isomorphic to the two-dimensional compact torus

$\mathbb{T}^2 = (\mathbb{R} \oplus \mathbb{R})/(\mathbb{Z} \oplus \mathbb{Z})$. Moreover, D_0 is a *cocompact* discrete subgroup of $\tilde{A}(\mathbb{R})$, i.e., the homogeneous manifold $D_0 \backslash \tilde{A}(\mathbb{R})$ of right cosets of D_0 in $\tilde{A}(\mathbb{R})$ is compact. Since $\tilde{C}/(D_0 \cap \tilde{C})$ can be identified with the compact torus $\mathbb{T} = \mathbb{R}/\mathbb{Z}$, the so-called compact *Heisenberg nilmanifold* $D_0 \backslash \tilde{A}(\mathbb{R})$ forms a *principal circle bundle over the two-dimensional torus* $\mathbb{T}^2$.

$$\begin{array}{ccc} \mathbb{T} & \longrightarrow & D_0 \backslash \tilde{A}(\mathbb{R}) \\ & & \downarrow \\ & & \mathbb{T}^2 \end{array}$$

Consider the continuous unitary character

$$\chi_1 : D \ni (\xi,\eta,z) \rightsquigarrow \chi_1(0,0,z) = e^{2\pi i z} \in \mathbb{T}$$

which forms the basic central character of D. The elements of the complex vector space $\mathbb{K}(D \backslash \tilde{A}(\mathbb{R}),\chi_1)$ can be identified with the *continuous* functions $\phi : \mathbb{R} \oplus \mathbb{R} \to \mathbb{C}$ satisfying the periodicity condition

$$\phi(x + \xi,\ y + \eta) = e^{2\pi i \xi y}\phi(x,y)$$

for $(x,y) \in \mathbb{R} \oplus \mathbb{R}$ and $(\xi,\eta) \in \mathbb{Z} \oplus \mathbb{Z}$. Then the elements of the complex vector space $\mathbb{K}(D_0 \backslash \tilde{A}(\mathbb{R}),\chi_1)$ can be identified with the continuous functions $\phi : \tilde{A}(\mathbb{R}) \to \mathbb{C}$ satisfying the condition

$$\phi(x + \xi,\ y + \eta,\ z + \zeta) = e^{2\pi i(\xi y + z)}\phi(x,y,0)$$

for $(x,y,z) \in \tilde{A}(\mathbb{R})$ and $(\xi,\eta,\zeta) \in D_0$. Consequently the functions $\phi \in \mathbb{K}(D_0 \backslash \tilde{A}(\mathbb{R}),\chi_1)$ are determined by their restrictions to the intersection of the polarized cross-section to $\tilde{C}$ in $\tilde{A}(\mathbb{R})$ with the fundamental domain $\{(x,y,z) \in \tilde{A}(\mathbb{R}) \mid (x,y,z) \in [-\frac{1}{2}, \frac{1}{2}[^3\}$ of the compact Heisenberg nilmanifold $D_0 \backslash \tilde{A}(\mathbb{R})$. Endow $D_0 \backslash \tilde{A}(\mathbb{R})$ with the unique probability measure which is invariant under the natural right action of $\tilde{A}(\mathbb{R})$ on $D_0 \backslash \tilde{A}(\mathbb{R})$. Then the completion $L^2(D_0 \backslash \tilde{A}(\mathbb{R}),\chi_1)$ of the complex prehilbert space $\mathbb{K}(D_0 \backslash \tilde{A}(\mathbb{R}),\chi_1)$ is isomorphic to the completion $L^2(D \backslash \tilde{A}(\mathbb{R}),\chi_1)$ of the complex prehilbert space $L^2(D \backslash \tilde{A}(\mathbb{R}),\chi_1)$ and we can form the unitarily induced linear representation

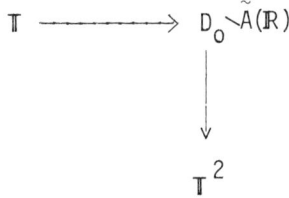

$$(U_{D_0}, L^2(D_0 \diagdown \tilde{A}(\mathbb{R}), \chi_1)) = \mathrm{Ind}_D^{\tilde{A}(\mathbb{R})}(\chi_1 . \mathrm{id}_{\mathbb{C}}, \mathbb{C})$$

of $\tilde{A}(\mathbb{R})$ which is called to be the linear *lattice representation* of $\tilde{A}(\mathbb{R})$. In view of the Stone-von Neumann-Mackey-Kinillov theorem (Theorem 7.4) we conclude

7.13 <u>Theorem</u>. The linear lattice representation $(U_{D_0}, L^2(D_0 \diagdown \tilde{A}(\mathbb{R}), \chi_1))$ of the real Heisenberg nilpotent Lie group $\tilde{A}(\mathbb{R})$ is isomorphic to the linear Schrödinger representation $(U, L^2(\mathbb{R}))$ of $\tilde{A}(\mathbb{R})$.

The last step is to make this isomorphy explicit. In order to avoid scaling factors it will be convenient to look at the dual pairing presentation of $\tilde{A}(\mathbb{R})$ and the realization $(U_1', L^2(\mathbb{R}))$ of the linear Schrödinger representation $(U, L^2(\mathbb{R}))$ of $\tilde{A}(\mathbb{R})$ introduced in 5.8. For $f \in \mathcal{S}(\mathbb{R})$ define

$$T_1 f(x,y,z) = \sum_{\xi \in \mathbf{Z}} (U_1(x,y,z)f)(\xi)$$

$$= e^{2\pi i z} \sum_{\xi \in \mathbf{Z}} e^{2\pi i \xi y} f(\xi + x)$$

for all elements $(x,y,z) \in \tilde{A}(\mathbb{R})$. Then $T_1 f \in \mathbb{K}(D_0 \diagdown \tilde{A}(\mathbb{R}), \chi_1)$ and $f \rightsquigarrow T_1 f$ extends to a norm-preserving $\mathbb{C}$-linear mapping of $L^2(\mathbb{R})$ into $L^2(D_0 \diagdown \tilde{A}(\mathbb{R}), \chi_1)$. On the other hand, let $\phi \in \mathbb{K}(D_0 \diagdown \tilde{A}(\mathbb{R}), \chi_1)$ and define

$$T_1^* : \mathbb{R} \ni x \rightsquigarrow \int_{\mathbb{T}} \phi(x,y,0)d\dot{y}$$

where $d\dot{y}$ denotes the normalized Haar measure of the torus group $\mathbb{R}/\mathbf{Z} = \mathbb{T}$. Then the complex numbers

$$T_1^*\phi(x + \xi) = \int_{\mathbb{T}} \phi(x,y,0)e^{2\pi i \xi y}d\dot{y} \qquad (\xi \in \mathbf{Z})$$

are the Fourier coefficients of the Fourier expansion of the function $\dot{y} \rightsquigarrow \phi(x,y,0)$ for $x \in \mathbb{R}$. Thus by the Parseval-Plancherel theorem we have

$$\int_{\mathbb{T}} |\phi(x,y,0)|^2 d\dot{y} = \sum_{\xi \in \mathbf{Z}} |T_1^*\phi(x + \xi)|^2$$

and therefore

$$\int_{\mathbb{R}} |T_1^*\phi(x)|^2 d\mu(x) = \int_{\mathbb{T}} \Big(\sum_{\xi \in \mathbb{Z}} |T_1^*\phi(x+\xi)|^2 d\dot{x}$$

$$= \int_{\mathbb{T}} \int_{\mathbb{T}} |\phi(x,y,0)|^2 d\dot{x}d\dot{y}.$$

Consequently T_1^* extends to a norm-preserving $\mathbb{C}$-linear mapping of $L^2(D_0\backslash\tilde{A}(\mathbb{R}),\chi_1)$ into $L^2(\mathbb{R})$ and for all functions $f \in \mathcal{S}(\mathbb{R})$, $\phi \in \mathcal{K}(D_0\backslash\tilde{A}(\mathbb{R}),\chi_1)$ we have the identity

$$\langle T_1 f|\phi\rangle = \langle f|T_1^*\phi\rangle.$$

From this we conclude that $T_1 : L^2(\mathbb{R}) \to L^2(D_0\backslash\tilde{A}(\mathbb{R}),\chi_1)$ is a unitary linear mapping such that

$$T_1^* = T_1^{-1}.$$

We will call T_1 the *Weil-Brezin isomorphism* of $L^2(\mathbb{R})$ onto $L^2(D_0\backslash\tilde{A}(\mathbb{R}),\chi_1)$. For all functions $f \in \mathcal{S}(\mathbb{R})$ and all elements $(x,y,z) \in \tilde{A}(\mathbb{R})$, $(x',y',z') \in \tilde{A}(\mathbb{R})$ we obtain

$$(T_1 \circ U_1'(x',y',z')f)(x,y,z) = e^{2\pi iz} \sum_{\xi \in \mathbb{Z}} e^{2\pi i\xi y} e^{2\pi i(z'+(x+\xi)y')} f(x+\xi+x')$$

$$= (U_{D_0}(x',y',z') \circ T_1(f)(x,y,z)$$

and therefore $T_1 \in \mathcal{R}_{\tilde{A}(\mathbb{R})}(U_1',U_{D_0})$. Consequently we established

7.14 <u>Theorem</u>. The Weil-Brezin isomorphism T_1 forms a unitary isomorphism of the topologically irreducible, continuous, unitary linear representation $(U_1',L^2(\mathbb{R}))$ of $\tilde{A}(\mathbb{R})$ onto the linear lattice representation $(U_{D_0},L^2(D_0\backslash\tilde{A}(\mathbb{R}),\chi_1))$ of $\tilde{A}(\mathbb{R})$.

In Section 7.7 we established the fundamental identity

$$h(\bar{\mathcal{F}}_{\mathbb{R}}) = j$$

for the Weyl element $j \in \underline{\underline{Sp}}(1,\mathbb{R})$. Let $J^\#$ denote the lifting of the automorphism $J = j \times \mathrm{id}_{\tilde{C}}$ of $\tilde{A}(\mathbb{R})$ to the complex vector space of all functions $\tilde{A}(\mathbb{R}) \to \mathbb{C}$. Then $J(D_0) = D_0$ and the Weil-Brezin isomorphism T_1 commutes with the Fourier cotransform $\bar{\mathcal{F}}_{\mathbb{R}}$ and $J^\#$ on $\mathcal{S}(\mathbb{R})$:

$$T_1 \circ \bar{\mathcal{F}}_R = J^\# \circ T_1.$$

We close this section by noting the following

Corollary. The Fourier cotransform of the real line $\mathbb{R}$ admits the factorization

$$\bar{\mathcal{F}}_{\mathbb{R}} = T_1^{-1} \circ J^\# \circ T_1$$

on $\mathcal{S}(\mathbb{R})$.

References

Auslander, L. : A factorization theorem for the Fourier transform of a separable locally compact abelian group. In: R.A. Askey, T.H. Koornwinder, and W. Schempp, editors: Special functions: Group theoretical aspects and applications, pp. 261-269. D. Reidel, Dordrecht-Boston-Lancaster 1984.

Bargmann, V. : On a Hilbert space of analytic functions and an associated integral transform I. Comm. Pure Appl. Math. 14 (1961), 187-214.

Segal, I. : Transforms for operators and symplectic automorphisms over a locally compact abelian group. Math. Scand. 13 (1963), 31-43.

Segal, I.E., Kunze, R.A. : Integrals and operators. Second edition. Springer-Verlag, Berlin-Heidelberg-New York 1978.

Weil, A. : Sur certains groupes d'opérateurs unitaires. Acta Math. 111 (1964), 143-211. Also in: Collected papers, Vol. III, pp. 1-69. Springer, New York-Heidelberg-Berlin 1980.

8 Applications to signal theory

8.1 Because radar computations are not familiar to the general mathematical community, let us begin with a brief explanation of the basic principles of radar in order to point out the rôle played by the three-dimensional real Heisenberg nilpotent Lie group $\tilde{A}(\mathbb{R})$ in the theory of analog radar signal design. To be more precise, we shall be concerned with an application of harmonic analysis on the differential principal fibre bundle over the two-dimensional polarized resp. isotropic cross-section with structure group isomorphic to the one-dimensional centre $\tilde{C}$ of the simply connected real Heisenberg nilpotent Lie group $\tilde{A}(\mathbb{R})$ to the mathematical treatment of the range and velocity measurement on one or more moving targets by means of the time delay and the Doppler frequency shift simultaneously attached to the analog signal return of a transmitted radar signal. It should be emphasized, however, that radar analysis and radar signal design are by no means the only applications of nilpotent harmonic analysis to signal theory. The dual reductive pair approach to physical phenomena involved in beam propagation, optical fibre and resonator techniques, and dielectric and metallic wave-guides appears to be another example of applications which are of great importance for terrestrial communication systems of high capacity.

The purpose of *radar* (= radio detection and ranging) systems is basically to survey broad areas of sky in order to detect the presence of distant objects and at the same time to gather various kinds of informations about the targets in the far field of the radar. In case of moving targets searched for by a radar the informations of main interest are the *bearing*, the *range* d, the *radial velocity* v relative to the antenna, and the *cross-section* of the objects.

The following figure shows an elementary form of a conventional radar system using a common stationary antenna ("radar dish") for both transmission and reception, which is achieved by means of a duplexer.

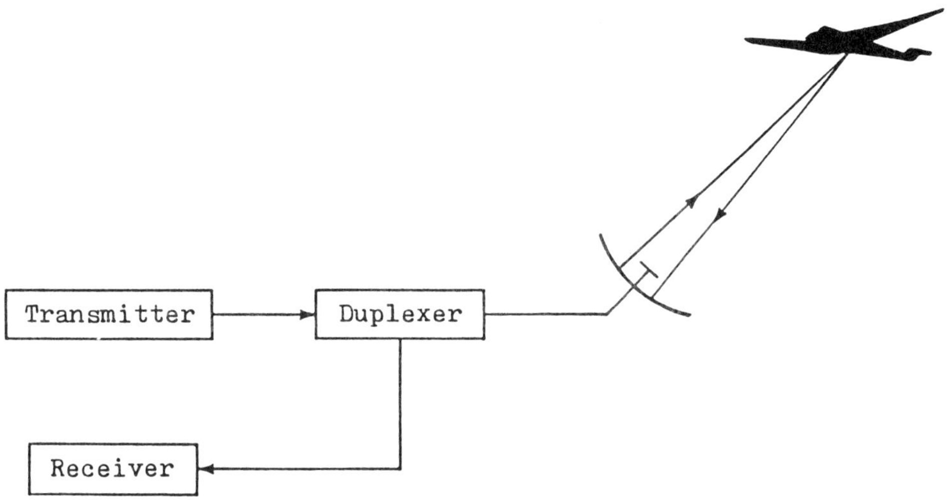

In a typical air traffic control radar the stationary antenna rotates mechanically, sweeping out a full 360° by its beam of microwave radiation every 4 to 12 sec. The azimuth beam resolution is about 1° to 2° and the vertical antenna pattern is a dispersed, fanlike beam, usually having a 30° to 45° width. Present airport surveillance radar systems must track up to 50 aircraft within their fields of view (typically 25 to 40 mi) and display these tracks to the air traffic controller. Yet in many of the most familiar uses of radar systems, such as aviation, air defense and intelligence, the mechanically steered parabolic antenna is giving way to a new kind of device. A flat, regular arrangement ("array antenna") of small, identical antenna elements, each one capable of transmitting and receiving radar signals, takes the place of the physically rotating parabolic reflector, and even as its beam scans expanses of sky no part of the radar antenna itself moves. Instead the signal is deflected from target to target electronically, steered through the physical principle of wave interference. Today a single radar system of this kind can do what previously might have required a battery of mechanically steered antennas. The new radar device is known as the electronically steerable *phased-array* system. In conjunction with high-speed digital signal processing hardware these new systems have led to great sophistication of radar tracking algorithms. However, the basic principles of radar still remain unchanged when embodied in this advanced technology.

In the transmission mode, the antenna radiates periodically a directed, narrow, pencil-like beam of radio energy in the form of trains and bursts of coherent radar pulses of large amplitude and brief duration and the same carrier frequency ω. A central oscillator generates the radar signal; transistors or specialized microwave tubes such as travelling-wave tubes amplify it. In order to estimate the targets' cross sections, bursts of extremely short pulses are necessary. Most often the electromagnetic energy falls within microwave bands, from 300 MHz to 10 GHz, although some very-long-range radar systems operate instead in the HF and VHF bands, respectively from 3 MHz to 30 MHz and from 30 MHz to 300 MHz. If some remote objects lie in the path of the propagating beam, a portion of the transmitted signal energy is reflected. Provided that multiple reflecting waves among the targets are negligible and the energy of the transmitted radar pulse, the sensitivity of the radar antenna (operating in the reception mode) and the reflective quality of the targets are all sufficient, a detectable echo will return to the antenna. The echo signal is then processed in the receiver to detect the presence of the targets and to gauge their parameters for precise tracking.

In the reception mode, the signals received by the radar antenna consist of a high-frequency carrier modulated in amplitude (or phase) by functions f of time that vary much more slowly than the cycles of the carrier. In radar, the parameters chiefly serving to distinguish or resolve two echo signals are their *arrival times* x and the *Doppler shifts* y of their carrier frequencies from a common reference frequency within the spectral band. The transmitted carrier frequency is a natural choice for the reference frequency of a single train of coherent pulses provided ω is very *stable*. Thus the analog part of the radar receiver assumes a high degree of accuracy.

The structure of the receiver and its performance depend upon the symmetrized *auto-correlation* or analog *radar auto-ambiguity function* $(x,y) \rightsquigarrow H(f;x,y)$. The function $H(f;\cdot,\cdot)$ associated with the *complex envelope* (or *time-varying phasor amplitude*) f of the monochromatic signal pulse

$$t \rightsquigarrow f(t)e^{2\pi i\omega t}$$

having the *carrier wave* $t \rightsquigarrow e^{2\pi i\omega t}$ and the analog signal return

$$t \rightsquigarrow f(t+x)e^{2\pi i\omega(t+y)}$$

that ideally is a time delayed and Doppler frequency shifted version of the transmitted signal will be explained in Section 8.2. For the purposes of present-day radars we may assume that the target range d depends linearly upon time during the transmission interval and that the target velocity v is small compared to the speed c of electromagnetic propagation (non-relativistic radar detection). It follows

$$d = \frac{1}{2} cx$$

and

$$v = \frac{1}{2} c \frac{y}{\omega} .$$

We always assume that the signal waveform f belongs to the Schwartz space $\mathcal{S}(\mathbb{R})$. By duality we may also include signals having tempered complex distributions $f \in \mathcal{S}'(\mathbb{R})$ as waveforms. We always consider $\mathcal{S}(\mathbb{R})$ as an (everywhere dense) vector subspace of the standard complex Hilbert space $L^2(\mathbb{R})$ which is embedded by the natural $\mathbb{C}$-linear mapping $f \rightsquigarrow f.\mu$ into $\mathcal{S}'(\mathbb{R})$. The energy of the signal is then given by the integral (squared L^2-norm)

$$\|f\|^2 = \int_{\mathbb{R}} |f(t)|^2 d\mu(t).$$

Thus the L^2-norm on the complex prehilbert space $\mathcal{S}(\mathbb{R})$ may be considered as the signal energy norm.

As an example of a radar signal which has attained practical importance we mention the *chirp signal* or *linear frequency modulated* (FM) *signal*. The complex envelope of the chirps takes the form

$$t \rightsquigarrow C_u e^{\pi i u t^2}$$

where $u \in \mathbb{R}$ and $C_u \in \mathbb{C}$ are non-zero constants. When trying first to detect a target at long range and then track the target as the range d decreases, a standard procedure is first to transmit a chirp signal followed by a pulse sequence, i.e., a burst yielding precise range and velocity measurements.

The practical realization of any radar system involves an enormous amount of complex electronic equipment. The brains of the radar system is the receiver tracking computer, scheduling the appropriate positions of the mechanically steered dish or the beam positions of the electronically steerable antenna, coordinating the transmitted radar signals via the control path going to the antenna system, and finally running the display system. The transmitted radar signals may vary from simple pulse trains to high bandwidth chirps and bursts of pulses and chirps. It is one of the most important tasks of the tracking computer to coordinate the transmitted radar signals via the control path going to the antenna system in order to achieve the best range and velocity measurement.

8.2 A major component in the design of the overall radar system is analog signal design which is directed toward achieving the best range and velocity measurement on one or more distant targets. At its centre is the analog *radar auto-ambiguity function*

$$(x,y) \rightsquigarrow H(f;x,y)$$

mentioned in Section 8.1 supra since it represents an idealized mathematical model of a radar system involving the two key variables, arrival time x and Doppler frequency shift y. It takes the symmetrized form

$$H(f;x,y) = \int_{\mathbb{R}} f(t + \tfrac{1}{2}x)\bar{f}(t - \tfrac{1}{2}x)e^{2\pi iyt}\, d\mu(t).$$

Since it is immaterial how the signs of time delay and Doppler frequency shift are chosen, there is no unique way of defining the radar auto-ambiguity function. Several essentially equivalent definitions exist in the literature.

While the radar auto-ambiguity function is complex-valued, the similarly defined *Wigner distribution function*

$$P(f;q,p) = \int_{\mathbb{R}} f(q + \tfrac{1}{2}t)\bar{f}(q - \tfrac{1}{2}t)e^{2\pi ipt}\, d\mu(t)$$

which governs the distribution of energy in the atomic nucleus is real-valued but not always positive on the position-momentum plane $\mathbb{R} \oplus \mathbb{R}$. Quantum mechanical phase space distribution functions have been studied extensively in quantum thermodynamics since this technique provides a useful tool for the treatment of quantum mechanical problems in terms of classical concepts. Apart of its use in quantum thermodynamics, the Wigner distribution $P(f;\cdot,\cdot)$ has important applications in the analysis of loudspeaker performance and first-order optical systems. The functions $H(f;\cdot,\cdot)$ and $P(f;\cdot,\cdot)$ are related through a double Fourier transform

$$P(f;\cdot,\cdot) = \mathcal{F}_{\mathbb{R}^2} H(f;\cdot,\cdot).$$

The symmetrized *cross-correlation* or analog *radar cross-ambiguity function* $H(f,g;\cdot,\cdot)$ associated with $f \in \mathcal{S}(\mathbb{R})$ and $g \in \mathcal{S}(\mathbb{R})$ is defined in an analogous way via the prescription

$$H(f,g;x,y) = \int_{\mathbb{R}} f(t + \tfrac{1}{2}x)\bar{g}(t - \tfrac{1}{2}x)e^{2\pi iyt}\, d\mu(t)$$

and is of importance in communication theory. Notice that the assignment

$$\mathcal{S}(\mathbb{R}) \times \mathcal{S}(\mathbb{R}) \ni (f,g) \rightsquigarrow H(f,g;\cdot,\cdot) \in \mathcal{S}(\mathbb{R} \oplus \mathbb{R})$$

defines a sesquilinear mapping which will be seen to be surjective. Its restriction $f \rightsquigarrow H(f;\cdot,\cdot)$ to the diagonal of $\mathcal{S}(\mathbb{R}) \times \mathcal{S}(\mathbb{R})$, however, is not surjective.

8.3 The first problem to be solved in analog radar signal design is the *synthesis problem*. It asks for an intrinsic characterization of those functions

$$F \in \mathcal{S}(\mathbb{R} \oplus \mathbb{R})$$

on the time-frequency plane $\mathbb{R} \oplus \mathbb{R}$ which belong to the range of the mapping $f \rightsquigarrow H(f;\cdot,\cdot)$, i.e., for which there exists a complex-valued envelope $f \in \mathcal{S}(\mathbb{R})$ satisfying the identity

$$F = H(f;\cdot,\cdot).$$

In other words, the problem is to find necessary and sufficient conditions for a given complex-valued smooth function F in the two Fourier dual variables $x \in \mathbb{R}$ (separation in time) and $y \in \mathbb{R}$ (separation in frequency) such that F can be realized as an analog radar auto-ambiguity function $(x,y) \rightsquigarrow H(f;x,y)$ with respect to a complex-valued smooth signal waveform $t \rightsquigarrow f(t)$

in one (time) variable $t \in \mathbb{R}$. In Theorem 8.10 infra we will establish a solution of the radar synthesis problem via harmonic analysis on the differential principal fibre bundle over the two-dimensional polarized resp. isotropic cross-section with structure group isomorphic to the one-dimensional centre $\tilde{C}$ of the simply connected real Heisenberg nilpotent Lie group $\tilde{A}(\mathbb{R})$. Ultimately, this approach to the synthesis problem is based on the analogy between non-relativistic quantum physics and signal theory.

8.4 The image $\mathcal{F} = H(f;\mathbb{R},\mathbb{R})$ of the time frequency plane $\mathbb{R} \oplus \mathbb{R}$ under the analog radar auto-ambiguity function $H(f;\cdot,\cdot)$ is called to be the *radar ambiguity surface* over the time-frequency plane generated by the complex envelope $f \in \mathcal{S}(\mathbb{R})$. For every signal the radar ambiguity surface is peaked at the origin $(0,0)$ of the time-frequency plane $\mathbb{R} \oplus \mathbb{R}$ so that certainly not all the functions $F \in \mathcal{S}(\mathbb{R} \oplus \mathbb{R})$ can be realized as analog radar auto-ambiguity functions with respect to a suitable signal waveform $f \in \mathcal{S}(\mathbb{R})$. A second signal arriving with separations x in time and y in frequency that lie under this central peak will be difficult to distinguish from the first signal. For many types of signals the radar ambiguity surface exhibits additional peaks elsewhere over the time-frequency plane. These sidelobes may conceal weak signals with arrival times and carrier frequencies far from those of the first signal. In a measurement of the arrival time and frequency of a single signal, the subsidiary peaks may lead to gross errors in the result. The taller the sidelobes of the radar ambiguity surface, the greater the probability of such errors in time and Doppler frequency shift. It is desirable, therefore, for the central peak of the radar ambiguity surface to be narrow, and for there to be as few and as low sidelobes as possible.

A transmitted narrow pulse results in good range but poor velocity measurement, while a wide pulse of a single frequency yields good velocity but bad range information. For instance, the chirp signal results in good range measurement but precise measurements of the velocity require additional waveforms such as pulse bursts. By changing the waveform f of a radar signal of given energy it is possible to change the accuracy of the range and relative radial velocity measurements in such a manner that an increase of the range accuracy results in a decrease of the velocity accuracy, and vice versa (range-velocity coupling). The basic constraint of analog radar signal design which has serious consequences on radar measurements, namely the fact that a radar signal cannot be designed that gives high performance everywhere in the range-velocity plane, constitutes the essence of the *radar uncertainty principle*. If the signal waveform $f \in \mathcal{S}(\mathbb{R})$ is normalized such that $\|f\| = 1$ holds, the radar uncertainty principle can be expressed in terms of the analog radar auto-ambiguity function by the formula

$$\iint_{\mathbb{R} \oplus \mathbb{R}} |H(f;x,y)|^2 d\mu(x)d\mu(y) = 1$$

which states that the total volume under the normalized radar ambiguity surface $\mathcal{F} = H(f;\mathbb{R},\mathbb{R})$ over the time-frequency plane equals unity, independent of the signal waveform f. It follows that there are bounds on achievable resolution performance in range d and radial velocity v so that radar signal design turns out to be a compromise between range and velocity measurement.

The radar uncertainty principle parallels the Heisenberg uncertainty principle of quantum mechanics although, at first thought, there appears to be no reason why the Heisenberg uncertainty principle should be of any consequence in radar theory. According to quantum mechanics not all the physical quantities observed in any realizable experiment (even in principle only) can be determined with an arbitrarily high accuracy. Even under ideal experimental conditions, an increase in the measurement accuracy of one quantity can be achieved only at the expense of decreasing the measurement accuracy on another "canonically conjugate" quantity. The position coordinate q and its momentum p is one example of two such canonically conjugate quantities: It is impossible to determine simultaneously the position q and momentum p of a non-relativistic quantum-mechanical particle (position-momentum coupling). If $f \in \mathcal{S}(\mathbb{R})$ denotes a normalized state vector, the identity

$$\iint_{\mathbb{R} \oplus \mathbb{R}} |P(f;q,p)|^2 \, d\mu(q)d\mu(p) = 1$$

is an expression for the Heisenberg uncertainty principle in terms of the Wigner distribution function.

8.5 To fully understand any mathematical system one has to understand the transformations of the system and especially those transformations of the system that leave some particular aspect of the system invariant. In case of the mathematical theory of analog radar signal design, a close investigation of the radar uncertainty principle leads to a study of the geometry of the radar ambiguity surfaces $\mathcal{F} = H(f;\mathbb{R},\mathbb{R})$ over the time-frequency plane by means of their energy preserving linear *automorphisms*. By such an automorphism of $\mathcal{F}$ we will understand a unitary operator

$$S : L^2(\mathbb{R}) \to L^2(\mathbb{R})$$

that maps the vector subspace $\mathcal{S}(\mathbb{R})$ onto itself such that for all waveforms $f \in \mathcal{S}(\mathbb{R})$ and for each pair $(x,y) \in \mathbb{R} \oplus \mathbb{R}$ there exists another pair $(x',y') \in \mathbb{R} \oplus \mathbb{R}$ depending on S that satisfies the identity

$$H(f;x,y) = H(S(f);x',y').$$

The second problem of analog radar signal design to be solved is the *invariant problem* for radar ambiguity surfaces over the time-frequency plane of calculating explicitly their energy preserving linear automorphisms. Indeed, the solution of the invariant problem which will be based on the linear oscillator representation (7.8) of the metaplectic group $\underline{\underline{Mp}}(1,\mathbb{R}) = \underline{\underline{Sp}}(1,\mathbb{R}) \times \{+1, -1\}$ and is given in Theorem 8.11 infra, is of fundamental importance for *computational signal geometry*. In particular, for every radar ambiguity surface over the time-frequency plane the result exhibits a generating procedure of the energy preserving linear automorphisms by means of chirp signals.

8.6 Recall from Section 5.9 that $(U_1', L^2(\mathbb{R}))$ is a version of the linear Schrödinger representation and therefore forms a topologically irreducible, continuous, unitary, linear representation of $\tilde{A}(\mathbb{R})$ in its basic presentation. Because $\tilde{A}(\mathbb{R})$ acts by $(U_1', L^2(\mathbb{R}))$ on the functions f of the Schwartz space $\mathcal{S}(\mathbb{R})$ according to the prescription

$$U_1'(x,y,z)f(t) = e^{2\pi i(z + yt + \frac{1}{2}xy)} f(t+x) \qquad (t \in \mathbb{R})$$

for all elements $(x,y,z) \in \tilde{A}(\mathbb{R})$, a change of variable yields

8.7 <u>Theorem</u>. The analog radar cross-ambiguity function associated with the signal envelopes $f \in \mathcal{S}(\mathbb{R})$ and $g \in \mathcal{S}(\mathbb{R})$ satisfies the identity

$$H(f,g;x,y) = c_{U_1',f,g}(x,y,0)$$

for all pairs $(x,y) \in \mathbb{R} \oplus \mathbb{R}$. Thus $H(f,g;\cdot,\cdot)$ equals the restriction of the coefficient function $c_{U_1',f,g}(\cdot,\cdot)$ of $(U_1', L^2(\mathbb{R}))$ onto the polarized cross-section to $\tilde{C}$ in $\tilde{A}(\mathbb{R})$.

From the basic identity of Theorem 8.7 evidence comes of the importance of harmonic analysis on the differential principal fibre bundle over cross-sections in $\tilde{A}(\mathbb{R})$ with structure group $\tilde{C}$ for the theory of analog radar signal design.

Since the linear representation $(U_1', L^2(\mathbb{R}))$ of $\tilde{A}(\mathbb{R})$ is square integrable modulo $\tilde{C}$ by Theorem 7.4, the orthogonality conditions of the coefficient functions $c_{U_1',f,g}$ imply (cf. Theorem 3.4) the following

Corollary (Moyal). The identity

$$\langle H(f,g;\cdot,\cdot) | H(f',g';\cdot,\cdot)\rangle = \langle f|f'\rangle \cdot \langle g'|g\rangle$$

holds for all signal envelopes f, f', g, g' in $\mathcal{S}(\mathbb{R})$.

In particular the radar uncertainty principle (cf. Section 8.4 supra) takes the form

$$\iint_{\mathbb{R}\oplus\mathbb{R}} |H(f;x,y)|^2 d\mu(x) d\mu(y) = \|f\|^4$$

in terms of the analog radar auto-ambiguity function $H(f;\cdot,\cdot)$ associated with $f \in \mathcal{S}(\mathbb{R})$.

8.8 Theorem. For any given function $F \in \mathcal{S}(\mathbb{R}\oplus\mathbb{R})$ there exist signal envelopes $f \in \mathcal{S}(\mathbb{R})$ and $g \in \mathcal{S}(\mathbb{R})$ such that the identity

$$F = H(f,g;\cdot,\cdot)$$

holds.

Proof. According to Section 7.3 the mapping U^2 defines a unitary isomorphism of the complex Hilbert space $L^2(\tilde{A}(\mathbb{R})/\tilde{C}, \chi_1; \mu\otimes\mu)$ onto the complex Hilbert space $\mathcal{L}_2(L^2(\mathbb{R}))$ of all Hilbert-Schmidt operators on $L^2(\mathbb{R})$. Its restriction U^1 takes $\mathcal{S}(\tilde{A}(\mathbb{R})/\tilde{C})$ onto the kernel operators with kernels in $\mathcal{S}(\mathbb{R}\oplus\mathbb{R})$, which are of trace class. It follows that every function $F \in \mathcal{S}(\tilde{A}(\mathbb{R})/\tilde{C})$ appears as a coefficient function of $(U_1', L^2(\mathbb{R}))$ with respect to a certain pair of functions $f \in \mathcal{S}(\mathbb{R})$ and $g \in \mathcal{S}(\mathbb{R})$. —

Thus any smooth function F on the time-frequency plane $\mathbb{R}\oplus\mathbb{R}$ can be realized as the analog radar cross-ambiguity function $H(f,g;\cdot,\cdot)$ associated with certain smooth waveforms f and g on the real line $\mathbb{R}$. More precisely, the sesquilinear mapping

$$\mathcal{S}(\mathbb{R}) \times \mathcal{S}(\mathbb{R}) \ni (f,g) \rightsquigarrow H(f,g;\cdot,\cdot) \in \mathcal{S}(\mathbb{R} \oplus \mathbb{R})$$

is surjective. In case of the analog radar auto-ambiguity function, however, i.e., by restricting the mapping $(f,g) \rightsquigarrow H(f,g;\cdot,\cdot)$ to the diagonal of $\mathcal{S}(\mathbb{R}) \times \mathcal{S}(\mathbb{R})$, the situation becomes different.

8.9 Recall from Section 7.1 that a cross-section W to the centre $\mathfrak{c}$ in the real Heisenberg Lie algebra $\mathfrak{n}$ defines a two-dimensional real symplectic vector space (W;B) which can be identified with the tangent space $\mathfrak{c}^o$ of the flat coadjoint orbit $\mathfrak{c}^o + T^* \in \mathfrak{n}^*/\tilde{A}(\mathbb{R})$ associated with the isomorphy class of $(U_1, L^2(\mathbb{R}))$ under the Kirillov correspondence. In the following we will identify the time-frequency plane $\mathbb{R} \oplus \mathbb{R}$ with the symplectic vector space (W;B).

A function $F: W \to \mathbb{C}$ is called to be of *positive type* on the two-dimensional real *symplectic* vector space (W;B) or c_1-*positive definite* on W if for all finite sequences of vectors $(v_j)_{1 \leq j \leq N}$ belonging to W the matrix

$$(c_1(v_j,v_k)F(v_j-v_k))_{\substack{1 \leq j \leq N \\ 1 \leq k \leq N}}$$

is a positive semidefinite Hermitean matrix. See Section 1.19 and Remark 2 of Section 1.2 where the 2-*cocycles* $c_\lambda (\lambda \neq 0)$ of W in $\mathbb{T}$ have been introduced. Theorem 1.20 yields some other characterizations of smooth c_1-positive definite functions $F \in \mathcal{S}(W)$ via the fibred convolution product of the complex algebra $\mathcal{S}(W)$ defined in Section 7.1.

A function $F: W \to \mathbb{C}$ is called to be a *minimal* continuous function of positive type on the symplectic vector space (W;B) if every decomposition

$$F = F_1 + F_2$$

of F into a sum of continuous functions F_1 and F_2 of positive type on (W;B) in the sense above implies

$$F_\alpha = \zeta_\alpha . F$$

with scalars $\zeta_\alpha \in \mathbb{C}$ $(\alpha \in \{1,2\})$ and $\zeta_1 + \zeta_2 = 1$; cf. Section 1.23. An application of Theorem 1.24 combined with Lemma 1.13 and Corollary 2 of Theorem

1.6 then yields by fibering over the two-dimensional isotropic cross-section $\exp_{\tilde{A}(\mathbb{R})} W$ to $\tilde{C}$ in $\tilde{A}(\mathbb{R})$ the following solution of the *radar synthesis problem*.

8.10 Theorem. Let the function $F \in \mathcal{S}(\mathbb{R} \oplus \mathbb{R})$ be given. There exists a complex signal envelope $f \in \mathcal{S}(\mathbb{R})$ such that

$$F = H(f;\cdot,\cdot)$$

holds if and only if F is a minimal function of positive type on the symplectic time-frequency plane $(W;B)$. In this case, the time-varying phasor amplitudes $f \in \mathcal{S}(\mathbb{R})$ which can be synthesized from F are determined uniquely up to a multiplicative complex constant of modulus 1.

Observe that the topologically irreducible, continuous, unitary, linear representation $(U_1', L^2(\mathbb{R}))$ of $\tilde{A}(\mathbb{R})$ is unitarily isomorphic to the linear Schrödinger representation $(U, L^2(\mathbb{R}))$ of $\tilde{A}(\mathbb{R})$. An application of Lemma 1.13 and Theorem 7.7 yields the following solution of the *invariant problem* for radar ambiguity surfaces over the symplectic time-frequency plane $(W;B)$ in terms of the linear oscillator representation $\tilde{\sigma} \rightsquigarrow T_{\tilde{\sigma}}$ (cf. Section 7.8) of the metaplectic group $\underline{\underline{Mp}}(1,\mathbb{R})$.

8.11 Theorem. Let the unitary operator

$$S : L^2(\mathbb{R}) \to L^2(\mathbb{R})$$

be an energy preserving linear automorphism of the radar ambiguity surface $\mathcal{F}$ over the symplectic time-frequency plane $(W;B)$ - then there exists a unique linear transformation $\sigma \in \underline{\underline{Sp}}(1,\mathbb{R})$ and a complex number $\zeta_{\tilde{\sigma}}$ of modulus $|\zeta_{\tilde{\sigma}}| = 1$ such that

$$S = \zeta_{\tilde{\sigma}} T_{\tilde{\sigma}}$$

holds.

Remark. In laser optics, the 2-cocycle associated with the projective Segal-Shale-Weil metaplectic representation $\sigma \rightsquigarrow T_\sigma$ of $\underline{\underline{Sp}}(1,\mathbb{R})$ gives rise to the Gouy phase-shift factor of optical resonator eigenmodes.

The Bruhat decomposition (cf. Section 7.7 supra)

$$\underline{\underline{Sp}}(1,\mathbb{R}) = NA \cup NAjN$$

where j denotes the Weyl element of $\underline{\underline{Sp}}(1,\mathbb{R})$ gives rise to the following generating procedure of the linear automorphisms of $\mathcal{F}$.

Corollary. The energy preserving linear automorphisms S of the radar ambiguity surface $\mathcal{F} = H(f;\mathbb{R},\mathbb{R})$ may be realized by finite sequences of time scalings, pointwise multiplications, and convolutions on the time axis $\mathbb{R}$ of the given signal waveform $f \in \mathcal{S}(\mathbb{R})$ with the chirp signal envelopes

$$t \rightsquigarrow C_b e^{\pi i b t^2} \quad (C_b \in \mathbb{C}, \quad |C_b| = 1)$$

with suitable real parameters $b \neq 0$.

The preceding result puts in evidence the distinguished rôle played by the time-varying chirp phasors in computational signal geometry.

8.12 Recall from Section 7.8 that the real Lie algebra of the metaplectic group $\underline{\underline{Mp}}(1,\mathbb{R})$ which is a twofold covering group of the symplectic group $\underline{\underline{Sp}}(1,\mathbb{R}) = \underline{\underline{SL}}(2,\mathbb{R})$, can be identified with $\mathfrak{sp}(1,\mathbb{R})$. Then the Weyl element

$$j = \begin{pmatrix} 0 & 1 \\ -1 & 0 \end{pmatrix}$$

belongs to $\mathfrak{sp}(1,\mathbb{R})$ and $\exp_{\underline{\underline{Mp}}(1,\mathbb{R})}(\pi r j)$ has as its projection $h(\exp_{\underline{\underline{Mp}}(1,\mathbb{R})}(\pi r j))$ in $\underline{\underline{Sp}}(1,\mathbb{R})$ the rotation

$$\begin{pmatrix} \cos 2\pi r & \sin 2\pi r \\ -\sin 2\pi r & \cos 2\pi r \end{pmatrix} \in \underline{\underline{SO}}(2,\mathbb{R})$$

for all $r \in \mathbb{R}$. Thus

$$\mathrm{Ker}(h \circ \exp_{\underline{\underline{Mp}}(1,\mathbb{R})} | \mathbb{R} j) = \mathbb{Z} j.$$

It follows by restricting the topologically irreducible, continuous, unitary, linear representation $(V, L^2(\mathbb{R}))$ of $\tilde{A}(\mathbb{R})$ (cf. Section 7.9) to the polarized cross-section to $\tilde{C}$ in $\tilde{A}(\mathbb{R})$

$$T_{\exp_{\underline{\underline{Mp}}(1,\mathbb{R})}(\pi r j)} \circ V(s(x,y),0) \circ T_{\exp_{\underline{\underline{Mp}}(1,\mathbb{R})}(-\pi r j)} =$$

$$V(s(x \cos 2\pi r + y \sin 2\pi r, -x \sin 2\pi r + y \cos 2\pi r), 0)$$

for all numbers $r \in \mathbb{R}$ and $s \in \mathbb{R}$. Taking $s = 1$, an application of Theorem 8.11 supra shows that the radar ambiguity surface $\mathcal{F} = H(f;\mathbb{R},\mathbb{R})$ generated by $f \in \mathcal{S}(\mathbb{R})$ over the symplectic time-frequency plane (W,B) is $\underline{\underline{SO}}(2,\mathbb{R})$-*invariant* (or *radial*) if and only if f is simultaneously an eigenfunction of the one-parameter subgroup

$$T_{\exp_{\underline{\underline{Mp}}(1,\mathbb{R})}(\pi\mathbb{R}j)}$$

of the unitary group $\underline{\underline{U}}(L^2(\mathbb{R}))$. Taking the derivative with respect to the parameter s at $s = 0$ we get the following identity valid on $\mathcal{S}(\mathbb{R})$ for all $r \in \mathbb{R}$:

$$T_{\exp_{\underline{\underline{Mp}}(1,\mathbb{R})}(\pi rj)} \circ (-x \frac{d}{dt} + 2\pi iyt) \circ T_{\exp_{\underline{\underline{Mp}}(1,\mathbb{R})}(-\pi rj)} =$$

$$-(x \cos 2\pi r + y \sin 2\pi r)\frac{d}{dt} + 2\pi i(y \cos 2\pi r - x \sin 2\pi r)t$$

In particular, if $x = \frac{1}{2}$ and $y = -\frac{1}{2}i$ we get for the creation operator A^+ the identity

$$T_{\exp_{\underline{\underline{Mp}}(1,\mathbb{R})}(\pi rj)} \circ A^+ \circ T_{\exp_{\underline{\underline{Mp}}(1,\mathbb{R})}(-\pi rj)} = e^{-2\pi ir}A^+$$

valid for all $r \in \mathbb{R}$. In the case $x = \frac{1}{2}$, $y = \frac{1}{2}i$ we get similarly for the annihilation operator A^-:

$$T_{\exp_{\underline{\underline{Mp}}(1,\mathbb{R})}(\pi rj)} \circ A^- \circ T_{\exp_{\underline{\underline{Mp}}(1,\mathbb{R})}(-\pi rj)} = e^{2\pi ir}A^-$$

Consequently we have

$$T_{\exp_{\underline{\underline{Mp}}(1,\mathbb{R})}(\pi rj)} \circ (A^+\circ A^- + A^-\circ A^+) \circ T_{\exp_{\underline{\underline{Mp}}(1,\mathbb{R})}(-\pi rj)} = A^+\circ A^- + A^-\circ A^+$$

for all $r \in \mathbb{R}$. A computation based on the factorization

$$j = \begin{pmatrix} 1 & 0 \\ -1 & 1 \end{pmatrix} \begin{pmatrix} 1 & 1 \\ 0 & 1 \end{pmatrix} \begin{pmatrix} 1 & 0 \\ -1 & 1 \end{pmatrix}$$

establishes that the *infinitesimal generator* of the one-parameter subgroup $T_{\exp_{\underline{\underline{Mp}}(1,\mathbb{R})}(\pi\mathbb{R}j)}$ of the unitary group $\underline{\underline{U}}(L^2(\mathbb{R}))$ is given by

$$A^+ \circ A^- + A^- \circ A^+ = \frac{d}{dr}\Big|_{r=0} T_{\exp_{\underline{\underline{Mp}}(1,\mathbb{R})}(\pi r j)}.$$

According to Section 7.10, the Hermite differential operator $-2(A^+ \circ A^- + A^- \circ A^+)$ with domain $\mathcal{S}(\mathbb{R})$ forms an essentially self-adjoint linear differential operator in $L^2(\mathbb{R})$ of second order with constant coefficients. Its closure admits a pure point spectrum. The eigenvalues are given by $\{2\pi(2m+1) \mid m \in \mathbb{N}\}$ and all these are simple eigenvalues. The associated normalized eigenfunctions are the Hermite functions $(h_m)_{m \geq 0}$ considered in Section 7.10. Thus we established the following result.

8.13 Theorem. Let the signal envelope $f \in \mathcal{S}(\mathbb{R})$ have energy norm $\|f\| = 1$. The bivariate analog radar ambiguity function $H(f;\cdot,\cdot)$ associated with f is $\underline{\underline{SO}}(2,\mathbb{R})$-invariant over the symplectic time-frequency plane (W,B) if and only if

$$f = \zeta_m h_m$$

for an integer $m \geq 0$ and a complex number ζ_m of modulus $|\zeta_m| = 1$.

Based on the preceding theorem it can be proved that the *diamond solvable Lie group*, i.e., the semi-direct product of $\tilde{A}(\mathbb{R})$ with $\mathbb{T}$ operates in a natural way on the *radial* radar ambiguity surfaces. However, instead of going into the details we observe that Theorem 8.11 supra enables us to determine the closely related analog radar ambiguity functions that are invariant under the cyclic group of order 4 operating on the time-frequency plane (W,B). From 7.7 we know that $h(\bar{\mathcal{F}}_{\mathbb{R}}) = j$ holds. The eigenvector spaces of the Fourier cotransform $\bar{\mathcal{F}}_{\mathbb{R}}: L^2(\mathbb{R}) \to L^2(\mathbb{R})$ associated with the four eigenvalues i^k are the *Hermite-Wiener spaces*

$$HW_k = \widehat{\bigoplus_{m \in \mathbb{N}}} \mathbb{C}.h_{4m+k} \quad (k \in \{0,1,2,3\}).$$

In view of Theorem 7.11 the direct sum decomposition

$$L^2(\mathbb{R}) = \bigoplus_{0 \leq k \leq 3} HW_k$$

holds. Therefore we get the following result which, of course, can be considerably refined by considering cyclic subgroups of $SO(2,\mathbb{R})$ of higher order than order 4.

Corollary. The analog radar ambiguity function $H(f;\cdot,\cdot)$ associated with the non-zero waveform $f \in \mathcal{S}(\mathbb{R})$ is invariant under quarterturns about the origin of the symplectic time-frequency plane (W,B) if and only if $f \in HW_k \subset \mathcal{S}(\mathbb{R})$ for a (unique) number $k \in \{0,1,2,3\}$.

8.14 In the next step we calculate explicitly the (radial) analog radar auto-ambiguity function $H(h_m;\cdot,\cdot)$ for $m \geqq 0$ in the real coordinates (p,q) with respect to the standard symplectic basis $\{X,Y\}$ of (W,B). Using the notations introduced in Section 7 we get for all pairs $(p,q) \in \mathbb{R} \oplus \mathbb{R}$ the identities

$$H(h_m;p,q) = c_{V,h_m,h_m}(\exp_{\tilde{A}(\mathbb{R})}(pX + qY))$$

$$= \langle e^{V(pX+qY)}(h_m) | h_m\rangle$$

$$= \frac{1}{\sqrt{\pi^m\ m!}} \langle e^{V(pX+qY)}(A^+)^m(h_0) | h_m\rangle$$

$$= \frac{1}{\sqrt{\pi^m\ m!}} \delta_W(Z^+)^m \langle e^{V(pX+qY)}(h_0) | h_m\rangle.$$

In Section 7.11 we established

$$e^{V(pX+qY)}(h_0) = e^{-(\pi/2)(p^2+q^2)} e^{(p+iq)A^+}(h_0).$$

Consequently

$$\langle e^{V(pX+qY)}(h_0) | h_m\rangle = \frac{1}{m!} e^{-(\pi/2)(p^2+q^2)}(p+iq)^m \langle (A^+)^m(h_0) | h_m\rangle$$

$$= \sqrt{\frac{\pi^m}{m!}}\ e^{-(\pi/2)(p^2+q^2)}(p+iq)^m.$$

Taking into account that

$$\delta_W(X) = \frac{\partial}{\partial p} + \pi i q,$$

$$\delta_W(Y) = \frac{\partial}{\partial q} - \pi i p,$$

holds we get the identities

$$\delta_W(Z^+) = \frac{1}{2}\left(\frac{\partial}{\partial p} - i\frac{\partial}{\partial q}\right) - \frac{\pi}{2}(p - iq)$$

$$= \frac{1}{2}e^{(\pi/2)(p^2+q^2)}\left(\frac{\partial}{\partial p} - i\frac{\partial}{\partial q}\right)e^{-(\pi/2)(p^2+q^2)} .$$

Let $L_m = L_m^{(0)}$ $(m \in \mathbb{N})$ denote the mth *Laguerre function* (of order 0) given by

$$L_m(x) = e^{-\frac{1}{2}x} \sum_{0 \leqq j \leqq m} \binom{m}{j} \frac{(-x)^j}{j!} \quad (x \in \mathbb{R})$$

then we get the final expression

$$H(h_m;p,q) = e^{-(\pi/2)(p^2+q^2)} \sum_{0 \leqq j \leqq m} \binom{m}{j} \frac{(-\pi(p^2+q^2))^j}{j!}$$

$$= L_m(\pi(p^2+q^2)) \quad (m \in \mathbb{N})$$

for all pairs $(p,q) \in \mathbb{R} \oplus \mathbb{R}$. Summarizing we established

8.15 Theorem. The (radial) analog radar auto-ambiguity functions $H(h_m;\cdot,\cdot)$ take for all integers $m \geqq 0$ the form

$$H(h_m;p,q) = L_m(\pi(p^2+q^2)) \quad (m \in \mathbb{N})$$

where $(p,q) \in \mathbb{R} \oplus \mathbb{R}$.

The following figures show the radial ambiguity surfaces $H(h_m;\mathbb{R},\mathbb{R})$ over the symplectic time-frequency plane $(W;B)$ in the cases $0 \leqq m \leqq 5$ (cf. Section 7.10).

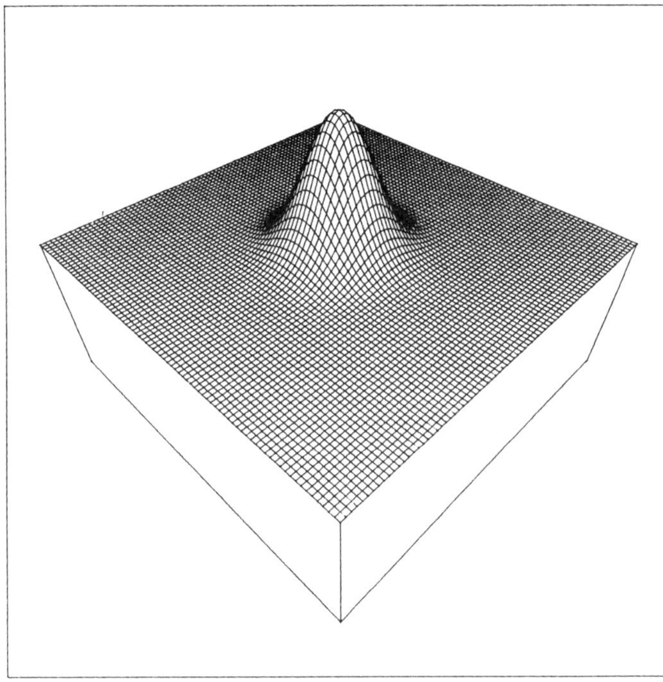

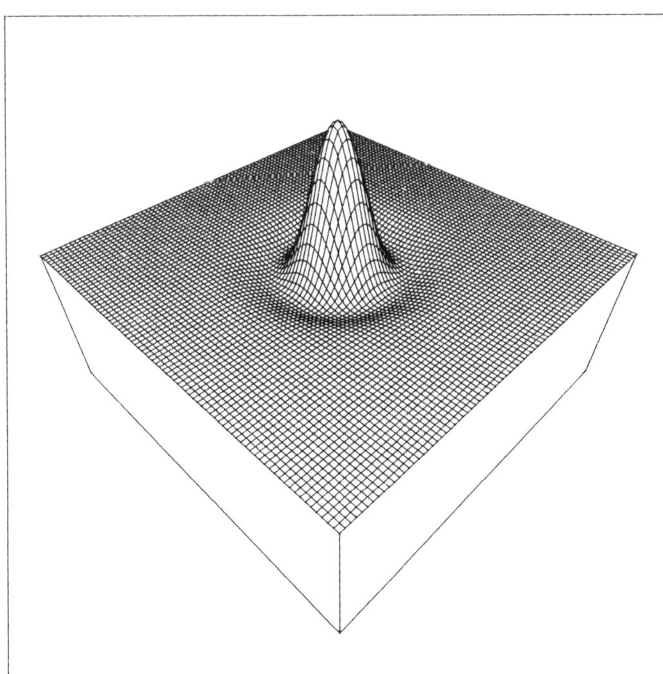

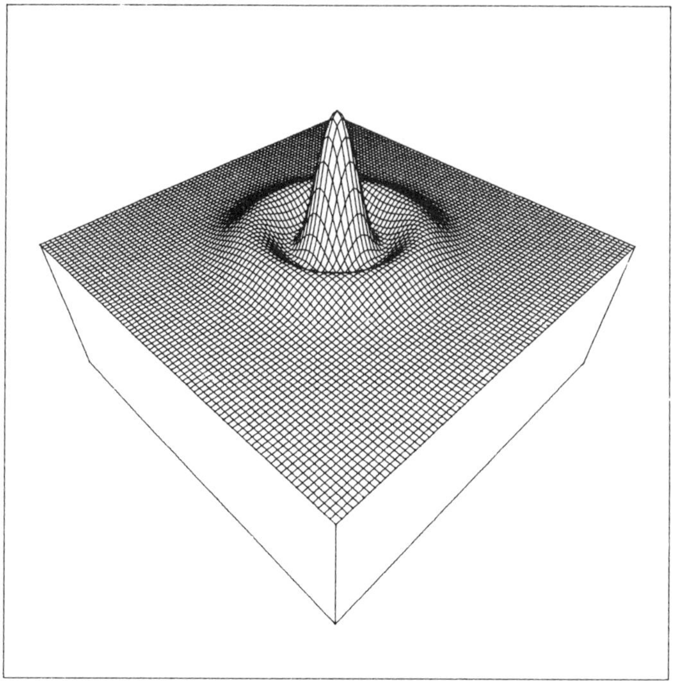

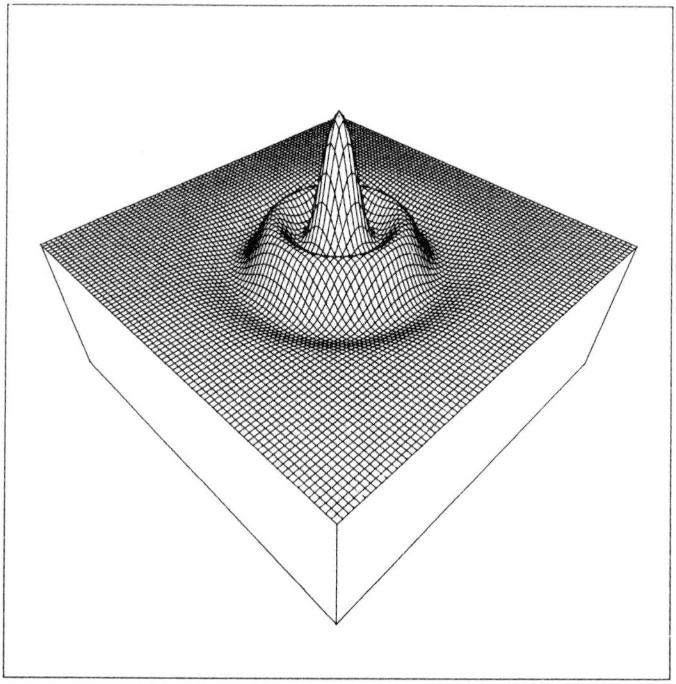

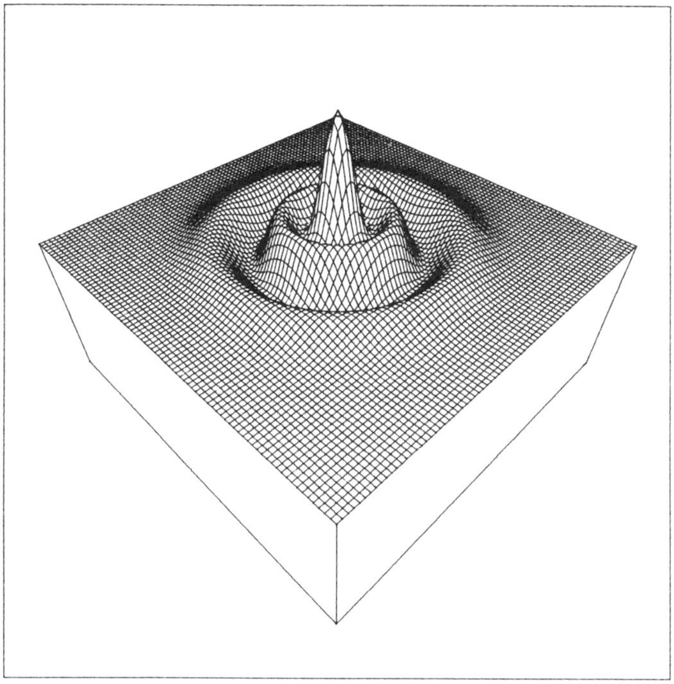

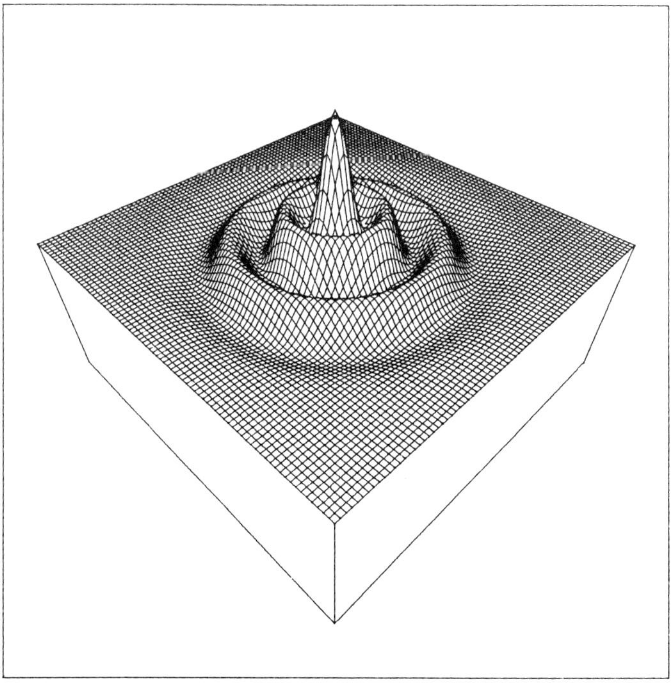

The calculations outlined in 8.14 may be extended to the analog radar cross-ambiguity functions $H(h_m,h_n;\cdot,\cdot)$. Indeed, denote by

$$L_n^{(\alpha)}(x) = e^{-\frac{1}{2}x} \sum_{0 \leq j \leq n} \binom{n+\alpha}{n-j} \frac{(-x)^j}{j!} \qquad (x \in \mathbb{R})$$

the nth *Laguerre function* of order $\alpha > -1$ where $n \geqq 0$ is an integer. Then we obtain for all integers $m \geqq n \geqq 0$ and all pairs $(p,q) \in \mathbb{R} \oplus \mathbb{R}$:

$$\begin{aligned}
H(h_m,h_n;p,q) &= c_{V,h_m,h_n}(\exp_{\tilde{A}(\mathbb{R})}(pX+qY)) \\
&= \langle e^{V(pX+qY)}(h_m)|h_n\rangle \\
&= \frac{1}{\sqrt{\pi^m\, m!}} \langle e^{V(pX+qY)}(A^+)^m(h_0)|h_n\rangle \\
&= \frac{1}{\sqrt{\pi^m\, m!}} \delta_W(Z^+)^m \langle e^{V(pX+qY)}(h_0)|h_n\rangle \\
&= \frac{1}{2^m\sqrt{\pi^{m-n} m!\; n!}} e^{(\pi/2)(p^2+q^2)}\left(\frac{\partial}{\partial p} - i\frac{\partial}{\partial q}\right)^m (p+iq)^n e^{-\pi(p^2+q^2)} \\
&= \sqrt{\frac{n!}{\pi^{m-n} m!}}\, e^{-(\pi/2)(p^2+q^2)} \sum_{0 \leqq j \leqq n} \binom{m}{j}\frac{1}{(n-j)!}(-\pi(p-iq))^{m-j}(p+iq)^{n-j} \\
&= \sqrt{\frac{n!}{m!}}\, e^{-(\pi/2)(p^2+q^2)}(-\sqrt{\pi}(p-iq))^{m-n} \sum_{0 \leqq j \leqq n} \binom{m}{j}\frac{1}{(n-j)!}(-(p^2+q^2))^{n-j}
\end{aligned}$$

By inserting the expression

$$L_n^{(m-n)}(x) = e^{-(1/2)x} \sum_{0 \leqq j \leqq n} \binom{m}{j} \frac{(-x)^{n-j}}{(n-j)!} \qquad (m \geqq n)$$

we get the following

8.16 Theorem. The analog radar cross-ambiguity functions with respect to the harmonic oscillator wave functions $(h_m)_{m \geqq 0}$ as signal envelopes take the form

$$H(h_m,h_n;p,q) = \sqrt{\frac{n!}{m!}}\,(\sqrt{\pi}\,(-p+iq))^{m-n} L_n^{(m-n)}(\pi(p^2+q^2)) \qquad (m \geqq n),$$

$$H(h_m,h_n;p,q) = \sqrt{\frac{m!}{n!}}\,(\sqrt{\pi}(p+iq))^{n-m} L_m^{(n-m)}(\pi(p^2+q^2)) \qquad (n \geqq m),$$

for all pairs $(p,q) \in \mathbb{R} \oplus \mathbb{R}$.

In laser optics, the squared expressions

$$|H(h_m,h_n;\cdot,\cdot)|^2 \qquad ((m,n) \in \mathbb{N} \times \mathbb{N})$$

may be interpreted as the *radial intensity distribution* of transverse Hermite-Gaussian eigenmodes of cylindrical optical resonators and waveguides; cf. Section 7.10 supra. The transverse intensity pattern appearing in the output beam of the laser takes the form

$$|H(h_m,h_n;\cdot,\cdot)|^2 \cos^2 2(m-n)\pi\theta \qquad ((m,n) \in \mathbb{N}\times\mathbb{N}\,,\ \theta \in [0,1[)$$

where θ denotes the azimuthal angle in a plane transverse to the beam direction. The following figures display various examples of transverse eigenmode patterns and some of their linear super-positions. It should be noted that the transverse eigenmodes govern the beam divergence, the beam diameter, and the energy density distribution in a plane perpendicular to the propagation of the light beam emitted from the laser. Therefore the identities calculated in Theorem 8.16 by means of nilpotent harmonic analysis methods play an important rôle in the theory of circularly symmetric optical resonators and distributed-index round waveguides. Their full group theoretical importance for beam and fibre optics, however, can be appreciated only when reductive dual pairs are considered.

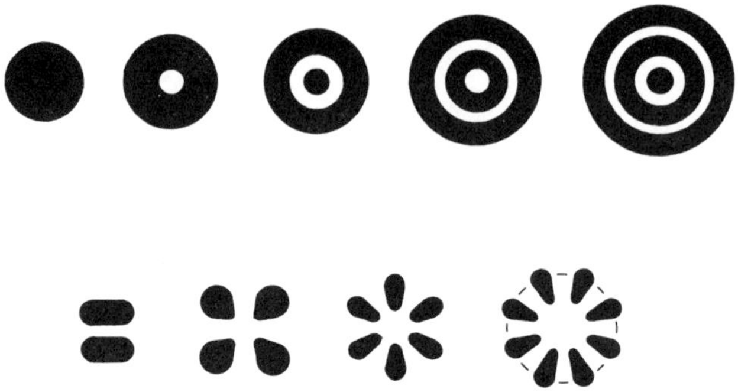

By restricting the identities displayed in Theorem 8.16 to the lattice $\mathbb{Z} \oplus \mathbb{Z}$ we conclude

<u>Corollary</u>. For all quadratic lattice points $(\mu,\nu) \in \mathbb{Z} \oplus \mathbb{Z}$ the identities

$$H(h_m,h_n;\mu,\nu) = \sqrt{\frac{n!}{m!}}\,(\sqrt{\pi}\,(-\mu+i\nu))^{m-n} L_n^{(m-n)}(\pi(\mu^2+\nu^2)) \qquad (m \geq n),$$

$$H(h_m,h_n;\mu,\nu) = \sqrt{\frac{m!}{n!}}\,(\sqrt{\pi}(\mu+i\nu))^{n-m}\, L_m^{(n-m)}(\pi(\mu^2+\nu^2)) \qquad (n \geq m),$$

hold.

8.17 Embed the quadratic lattice $\mathbb{Z} \oplus \mathbb{Z}$ into the discrete cocompact subgroup D_0 of $\tilde{A}(\mathbb{R})$ in the natural way:

$$(\mu,\nu) \rightsquigarrow (\mu,\nu,0).$$

Recall from Theorem 7.14 supra that the Weil-Brezin isomorphism T_1 forms a unitary isomorphism of the topologically irreducible, continuous, unitary, linear representation $(U_1',L^2(\mathbb{R}))$ of $\tilde{A}(\mathbb{R})$ in its dual pairing presentation onto the linear lattice representation $(U_{D_0},L^2(D_0\backslash\tilde{A}(\mathbb{R}),\chi_1))$. Let T_1 denote the Weil-Brezin isomorphism implemented by the basic presentation of $\tilde{A}(\mathbb{R})$. In view of Theorem 8.7 we then get

8.18 <u>Theorem</u>. For all signal envelopes $f \in \mathcal{S}(\mathbb{R})$ and $g \in \mathcal{S}(\mathbb{R})$ the associated analog radar cross-ambiguity function admits the form

$$H(f,g;x,y) = c_{U_{D_0},T_1(f),T_1(g)}(x,y,0)$$

where $(x,y) \in \mathbb{R} \oplus \mathbb{R}$. Thus $H(f,g;\cdot,\cdot)$ equals the restriction of the coefficient function $c_{U_{D_0},T_1(f),T_1(g)}(\cdot,\cdot,\cdot)$ of $(U_{D_0}, L^2(D_0 \setminus \tilde{A}(\mathbb{R}))$, onto the polarized cross-section to $\tilde{C}$ in $\tilde{A}(\mathbb{R})$.

Retain the above notations - then an application of the Fourier inversion for doubly periodic infinitely differentiable functions yields the following result.

<u>Corollary</u>. For any two functions f and g in $\mathcal{S}(\mathbb{R})$ the absolutely convergent Fourier expansion

$$(T_1(f).T_1(g))(x,y,0) = \sum_{(\mu,\nu) \in \mathbb{Z} \oplus \mathbb{Z}} H(f,g;\mu,\nu)e^{2\pi i(-\nu x+\mu y)}$$

holds for all points $(x,y) \in \mathbb{R} \oplus \mathbb{R}$ with respect to the topology of uniform $\mathfrak{C}^{\infty}$-convergence.

In view of the preceding identity, the Parseval-Plancherel theorem for double Fourier series, to wit, the canonical unitary isomorphy between the complex Hilbert spaces $L^2(\mathbb{T}^2)$ and $L^2(\mathbb{Z}^2)$ yields the following identity for the evaluations of analog radar ambiguity functions at quadratic lattice points.

8.19 <u>Theorem</u>. For all functions $f \in \mathcal{S}(\mathbb{R})$ and $g \in \mathcal{S}(\mathbb{R})$ the identity

$$\sum_{(\mu,\nu) \in \mathbb{Z} \oplus \mathbb{Z}} H(f;\mu,\nu).H(g;\mu,\nu) = \sum_{(\mu,\nu) \in \mathbb{Z} \oplus \mathbb{Z}} |H(f,g;\mu,\nu)|^2$$

holds.

In the radial case we get by the Corollary of Theorem 8.16 the following

<u>Corollary</u>. For all integers $m \geq n \geq 0$ the following identity for Laguerre functions of different orders holds:

$$\sum_{(\mu,\nu)\in\mathbb{Z}\times\mathbb{Z}} L_m^{(0)}(\pi(\mu^2+\nu^2)).L_n^{(0)}(\pi(\mu^2+\nu^2))$$

$$= \frac{n!}{m!}\pi^{m-n} \sum_{(\mu,\nu)\in\mathbb{Z}\times\mathbb{Z}} (\mu^2+\nu^2)^{m-n}(L_n^{(m-n)}(\pi(\mu^2+\nu^2)))^2.$$

In the case m = 1, n = 0 we get the identity

$$\pi \sum_{\mu\in\mathbb{Z}} \mu^2 e^{-\pi\mu^2} = \frac{1}{4}\sum_{\mu\in\mathbb{Z}} e^{-\pi\mu^2}.$$

The case m = 2, n = 1 yields the identity

$$\pi^3 \sum_{\mu\in\mathbb{Z}} \mu^6 e^{-\pi\mu^2} = \frac{15}{32}\sum_{\mu\in\mathbb{Z}} (8\pi^2\mu^4 - 1)e^{-\pi\mu^2}.$$

Moreover, the case m = 3, n = 2 implies

$$\pi^5 \sum_{\mu\in\mathbb{Z}} \mu^{10} e^{-\pi\mu^2} = \frac{45}{64}\sum_{\mu\in\mathbb{Z}} (16\pi^4\mu^8 - 140\pi^2\mu^4 + 21)e^{-\pi\mu^2}.$$

Finally, the case m = 4, n = 3 gives

$$\pi^7 \sum_{\mu\in\mathbb{Z}} \mu^{14} e^{-\pi\mu^2} = \frac{91}{1024}\sum_{\mu\in\mathbb{Z}} (256\,\pi^6\mu^{12} - 15840\,\pi^4\mu^8 + 166320\,\pi^2\mu^4 - 25245)e^{-\pi\mu^2}.$$

There is another proof of the preceding identities for the *theta constants* (or *theta-null values*)

$$\theta(1) = \theta(0,1) = \sum_{\mu\in\mathbb{Z}} e^{-\pi\mu^2}$$

by performing the derivatives of the transformation formula of the classical Jacobi theta function θ.

8.20 Let M_1 denote the indicator function of one side of the box $[-\frac{1}{2},+\frac{1}{2}[^3$ which forms a fundamental domain of the compact Heisenberg nilmanifold $D_0\backslash\tilde{A}(\mathbb{R})$. Thus

$$M_1(x) = \begin{cases} 1 & \text{for } x \in [-\frac{1}{2}, +\frac{1}{2}[, \\ 0 & \text{otherwise,} \end{cases}$$

i.e., M_1 forms the *basis spline function* of degree 0 on $\mathbb{R}$. Let $g \in L^1(\mathbb{R})$ be given - then

$$\sum_{n \in \mathbb{Z}} \int_{\mathbb{R}} |g(n+x)| M_1(x) d\mu(x) = \|g\|_1 .$$

Thus the infinite series $T_1(g)(x,0,0)$ converges for μ-almost all points $x \in \mathbb{R}$ towards a complex-valued function which is absolutely integrable over the compact torus group $\mathbb{T}$. Its equivalence class may be considered as an element of $L^1(\mathbb{T})$ and its Fourier coefficients can be computed by the dominated convergence theorem as follows:

$$\sum_{n \in \mathbb{Z}} \int_{\mathbb{R}} g(n+x) M_1(x) e^{-2\pi ikx} d\mu(x) = \mathcal{F}_{\mathbb{R}}\, g(k) = \bar{\mathcal{F}}_{\mathbb{R}}\, g(-k) \qquad (k \in \mathbb{Z}).$$

An application of the factorization of the Fourier contransform $\mathcal{F}_{\mathbb{R}}$ by means of the Weil-Brezin isomorphism T_1 as established in the Corollary of Theorem 7.14 supra shows that

$$\sum_{n \in \mathbb{Z}} \bar{\mathcal{F}}_{\mathbb{R}}\, g(n) e^{-2\pi inx}$$

represents the Fourier series of $T_1 g(x,0,0)$ and that for all $y \in \mathbb{R}$ the identity

$$\sum_{n \in \mathbb{Z}} \int_{\mathbb{R}} g(n+x) M_1(x) e^{2\pi iyx} d\mu(x) = \sum_{n \in \mathbb{Z}} \bar{\mathcal{F}}_{\mathbb{R}}\, g(n) \int_{-1/2}^{+1/2} e^{2\pi i(y-n)x} d\mu(x)$$

$$= \sum_{n \in \mathbb{Z}} \bar{\mathcal{F}}_{\mathbb{R}}\, g(n) \frac{\sin \pi(y-n)}{\pi(y-n)}$$

holds. Introduce the entire holomorphic function sinc ("*sinus cardinalis*") according to the prescription

$$\text{sinc}: \mathbb{C} \ni w \rightsquigarrow \begin{cases} \dfrac{\sin \pi w}{\pi w} & \text{for } w \neq 0 \\ 1 & \text{otherwise.} \end{cases}$$

Then we have

$$\sum_{n \in \mathbb{Z}} \int_{n-\frac{1}{2}}^{n+\frac{1}{2}} g(x)e^{2\pi iy(x-n)}d\mu(x) = \sum_{n \in \mathbb{Z}} \bar{\mathcal{F}}_{\mathbb{R}}\, g(n)\operatorname{sinc}(y-n)$$

for all $y \in \mathbb{R}$. Hence

$$\sum_{n \in \mathbb{Z}} (1-e^{-2\pi iny}) \int_{n-\frac{1}{2}}^{n+\frac{1}{2}} g(x)e^{2\pi iyx}d\mu(x) = \bar{\mathcal{F}}_{\mathbb{R}}\, g(y) - \sum_{n\in\mathbb{Z}} \bar{\mathcal{F}}_{\mathbb{R}}\, g(n)\operatorname{sinc}(y-n).$$

Provided $g(x) = 0$ for all points $x \in [-\frac{1}{2}, +\frac{1}{2}[$ on the line $\mathbb{R}$, we get the identity

$$\bar{\mathcal{F}}_{\mathbb{R}} g(y) = \sum_{n \in \mathbb{Z}} \bar{\mathcal{F}}_{\mathbb{R}}\, g(n)\operatorname{sinc}(y-n) \qquad (y \in \mathbb{R}).$$

An application of the Paley-Wiener theorem to the *band-limited* function $f = \bar{\mathcal{F}}_{\mathbb{R}}\, g$ yields the so-called *sampling theorem*.

8.21 Theorem (Whittaker-Shannon-Kotel'nikov). Let f denote an entire holomorphic function of exponential type $\leq \pi$ such that its restriction onto the real line $\mathbb{R}$ belongs to the complex Hilbert space $L^2(\mathbb{R})$. Then the cardinal series expansion

$$f(w) = \sum_{n \in \mathbb{Z}} f(n)\operatorname{sinc}(w-n)$$

holds for all $w \in \mathbb{C}$. The cardinal interpolation series converges uniformly on the compact subsets of $\mathbb{C}$.

The preceding result is a special case of a more general theorem dealing with *cardinal spline interpolation*. Indeed, the technique of factorizing the Fourier cotransform $\bar{\mathcal{F}}$ by means of the Weil-Brezin isomorphism T_1 as established in the Corollary of Theorem 7.14 supra can be applied to the convolution powers M_1^{*n} of the basis spline function M_1 of degree 0 on $\mathbb{R}$. The functions M_1^{*n} $(n \geq 0)$ form the basis splines of degree n on $\mathbb{R}$. An application of the Toeplitz-Wiener-Gel'fand theorem concerning the invertability within the commutative complex Banach algebra $\bar{\mathcal{F}}_{\mathbb{T}}L^1(\mathbb{Z})$ of (continuous) functions on $\mathbb{T}$ having an absolutely convergent Fourier series then yields the existence and uniqueness theorem of cardinal spline interpolation due to Subbotin and Schoenberg.

8.22 Theorem 8.21 supra may be interpreted in two ways, each of which has found important applications in signal theory.

(1) Every signal of finite energy and bandwidth W = 1/2 Hz may be completely recaptured, in a simple way, from a knowledge of its samples taken at the rate of 2W = 1 per second (*Nyquist rate*). Moreover - indispensable for any implementation in practice - the recovery is stable, in the sense that a small error in reading the sample values produces only a correspondingly small error in the recaptured signal.

(2) Every square-summable sequence of complex numbers may be transmitted at the rate of 2W = 1 per second over an ideal channel of bandwidth W = 1/2 Hz, by being represented as the samples at the integer points $n \in \mathbb{Z}$ of an easily constructed band-limited signal of finite energy.

Thus the Whittaker-Shannon-Kotel'nikov sampling theorem as stated in 8.21 serves as a basis for the interchangeability of analog representations of signals and their representations in digital sequences. Coding of a signal usually consists in its representation in a digital sequence. The digital form of representation rather than the analog representation of signals provides considerably more ways of protecting the signal against various kinds of possible distortion in its retaining and transmission. This is one reason among various other attendant advantages why the digital signal processing and therefore the Whittaker-Shannon-Kotel'nikov sampling theorem are so extremely valuable for modern communication systems of high capacity. The recently developed CD (= Compact Disc) technique and the glass fibres in optical communication systems which form excellent channels for the low-loss transmission of trains of optical pulses, are very efficient practical applications of the digital signal representation. These optoelectronic devices have become one of the most promising approaches to terrestrial communication since the laser first appeared. The achievement of optic technology has made the CD and the optical fibre the leading contenders as the storage and transmission media for a vast variety of current and future communication systems.

References

Brookner, E. : Radar technology. Artech House, Dedham, MA 1985.

Higgins, J.R. : Five short stories about the cardinal series. Bull. (New Series) Amer. Math. Soc. 12 (1985), 45-89.

Schempp, W. : Gruppentheoretische Aspekte der Signalübertragung und der kardinalen Interpolationssplines I. Math. Meth. in the Appl. Sci. 5 (1983), 195-215.

Schempp, W., Delvos, F.J. : Gruppentheoretische Aspekte der Signalübertragung und der kardinalen Interpolationssplines II (to appear).

Schempp, W. : Radar ambiguity functions, the Heisenberg group, and holomorphic theta series. Proc. Amer. Math. Soc. 92 (1984), 103-110.

Schempp, W. : Drei statt einer reellen Variablen? In: Multivariate approximation theory II. W. Schempp and K. Zeller, eds., pp. 331-341. ISNM 61. Birkhäuser, Basel-Boston-Stuttgart 1982.

Schempp, W. : Analog radar signal design and digital signal processing - a Heisenberg nilpotent Lie group approach. In: Lie methods in optics, Chapter 1. J. Sánchez-Mandragón and K.B. Wolf, eds. Lecture Notes in Physics. Springer, Berlin-Heidelberg-New York-Tokyo (in print).

Schempp, W. : The oscillator representation and laser optics I: The diamond solvable Lie group. C.R. Math. Rep. Acad. Sci. Canada (to appear).

Schempp, W. : The oscillator representation and laser optics II: The coupling of transverse modes. C.R. Math. Rep. Acad. Sci. Canada (to appear).

Schempp, W. : Signal geometry (in preparation).

Woodward, P.M. : Probability and information theory, with applications to radar. Artech House, Dedham, MA 1980.

Index